Life of Fred®

Decimals and Percents

Life of Fred®

Decimals and Percents

Stanley F. Schmidt, Ph.D.

Polka Dot Publishing

ISBN: 978-0-9791072-0-7

Library of Congress Catalog Number: 2007920296
Printed and bound in the United States of America

Polka Dot Publishing Reno, Nevada

To order copies of books in the Life of Fred series,

visit our Web site PolkaDotPublishing.com

Questions or comments? Email the author at lifeoffred@yahoo.com

Eleventh printing

for Goodness' sake

or as J.S. Bach—who was
never noted for his plain
English—often expressed it:

Ad Majorem Dei Gloriam

(to the greater glory of God)

A Note to Students

This is the last arithmetic book. After you finish the 208 pages of this book you will be ready for pre-algebra. There are 33 chapters. Each chapter is a lesson. Just like in *Life of Fred: Fractions,* after each five chapters you will come to **The Bridge**, which will give you a chance to show that you know the math before you move on to the next chapter.

The main danger in the *Life of Fred* books is that the readers enjoy them too much. The temptation is to zzzzoooooooooommmm through the chapters reading about the adventures of Fred. Here is a secret:

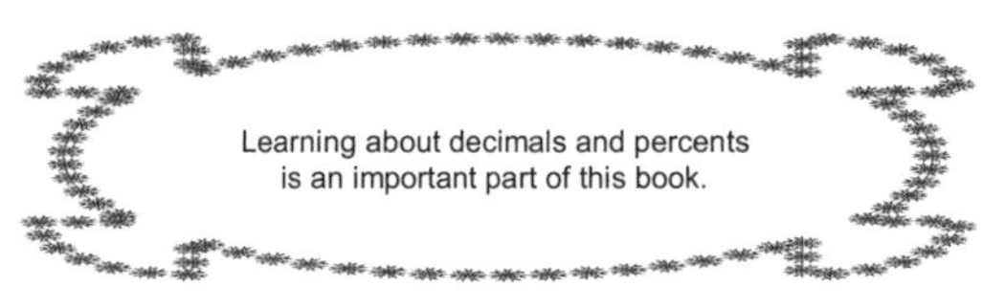

We continue our story of Fred where we left off at the end of *Life of Fred: Fractions.* As before, when I am writing I will use Times New Roman typeface. When Fred is thinking, he'll use this typeface. And when you, my reader, voice your questions (or complaints), ***you will use this typeface.***

Now that we have settled all of that, feel free to skip the rest of what is called the "front matter" and turn to page 13 to find out what Fred did after he opened the box that *didn't* contain his bicycle.

A Note to Parents

Mary Poppins was right: A spoonful of sugar can make life a little more pleasant. It is surprising that so few arithmetic books have figured that out.

Some arithmetic books omit the sugar—which is like lemonade without any sweetener. They give you a couple of examples followed by a zillion identical problems to do. And they call that a lesson. No wonder students aren't eager to read those books.

At the other extreme are the books that are just pure sugar—imagine a glass of lemonade with so much sugar in it that your spoon floats. The pages are filled with color and happy little pictures to show you how wonderful arithmetic is. The book comes with ① a teachers' manual, ② a computer disc, ③ a test booklet, and ④ a box of manipulatives. And they are so busy entertaining the reader that they don't teach a lot of math. This second approach is also usually quite expen$ive.

We'll take the Goldilocks approach: not too sour and not too sweet. We will also include a lot of mathematics. (Check out the Contents on page 10.) How many arithmetic books include both forms of the Goldbach conjecture? (See Chapter 17.) The reader will be ready for pre-algebra after completing this book.

This book covers one afternoon and evening of Fred's life and continues the story from *Life of Fred: Fractions.* Every piece of math *first* happens in his life, and then we do the math. It is all motivated by real life. When is the last time you saw prime numbers actually *used* in everyday life? They are needed in this book when the cavalry is getting ready to attack what the newspaper calls the "Death Monster."

FACTS ABOUT THE BOOK

Each chapter is a lesson. Thirty-three chapters = 33 lessons.

At the end of each chapter is a *Your Turn to Play*, which gives the student an opportunity to work with the material just presented. The answers are all supplied. The questions are not all look-alike questions. Some of them require . . . thought! *Your Turn to Play* often incorporates some review material. The students will get plenty of opportunity to keep using the material they have learned.

At the end of every five chapters is **The Bridge**, ten questions reviewing everything learned up to that point. If students want to get on to the next chapter, they need to show *mastery* of what has been covered so far. They need to get nine or more questions correct* in order to move on to the new material. If they don't succeed on the first try, there is a second set of ten questions—a second try. And a third try. And a fourth try. And a fifth try. Lots of chances to cross the bridge.

Don't let your students move on without showing mastery of the previous math.

At the end of the book is **The Final Bridge**, consisting of twenty questions. Again, five tries are offered.

RULES OF THE GAME

For now, students should put aside their calculators. This is the last chance we have to cement in place their addition and multiplication facts (which they should have had memorized before they began *Life of Fred: Fractions*). I balance my checkbook each month without a calculator just to keep in practice.

banned for now

Once the students get to pre-algebra they can take their calculators out of their drawers and use them all they like.

When the students are working on the *Your Turn to Play* or **The Bridge** sections, *they should write out their answers.* When they are working on a **Bridge**, they should complete the whole quiz first.

Then you and your child can check the answers together. This will give you a chance to monitor their progress. Mastery of the material is much more important than speed.

---

* The answers to all of the Bridge questions are given on pages 172–183.

FINAL THOUGHTS

Life of Fred books are designed to teach the material. They are not merely repositories of examples and homework problems. It is so important that kids

learn
how to learn
from reading.

Once they finish college, they will face forty or more years in which virtually all of their real learning* will come from *what they read.* It is not a favor to the students for you to repeat what the book said. If you do that, it is a disincentive for them to learn from their reading.

As strange as it sounds, *you don't need to teach the material.* I've done that work for you. Relax. You can best teach by example. You read your books while they read theirs.

The best way for you to help is to check their progress when they work on **The Bridges**.

✶ If "real learning" for adults is exemplified by what they see on television—on quiz shows or the educational channels—then the thousands of dollars and the thousands of hours they spent going to college were an utter waste.

Contents

Chapter One
Number Systems

What do five-and-a-half-year-old boys dream about? Many things. For Fred it was a new bicycle. When the box arrived at his office, he tore off the tape. The box fell open. Inside was . . . junk. There were gears, wires, rods, and motors, but no bicycle. He had spent every penny in his checking account ($1,935.06) and didn't get a bike.

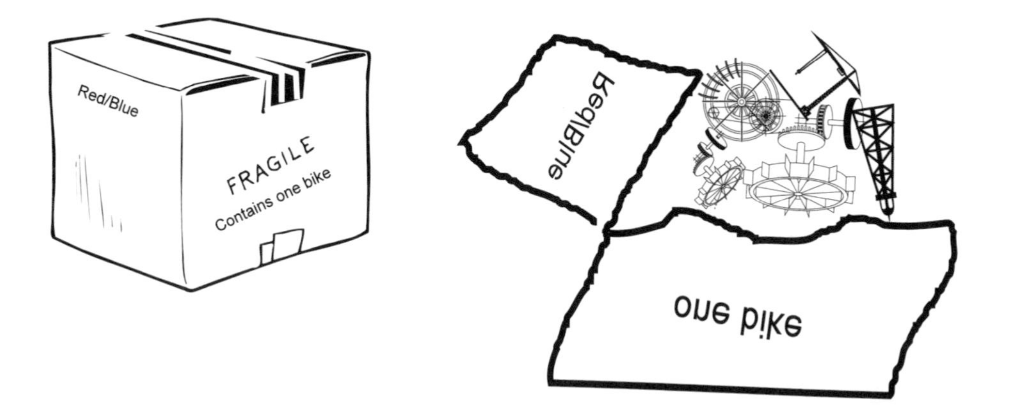

Fred had been cheated.

After a short trip with a blanket to a corner of his office to do a little crying, he returned to look at the pile of parts on the floor. There were bags of electrical plugs. There were springs. Fred thought *What shall I do with all this stuff? Maybe I should just throw it all in the garbage.*

Then he almost stepped on a huge remote control. It had about 168 buttons on it. And then it came to him: *I know! I will build a robot!*

Maybe my $1,935.06 won't be wasted after all.

Let's look at $1,935.06 for a moment.

1,935.06

This is a decimal number. That's because it contains a decimal point (the dot between the 5 and the 0).

When you studied the whole numbers, {0, 1, 2, 3, 4, . . . }, you didn't need any decimal points. When you count the number of buttons on a remote control, you get 168, not 168 $\frac{3}{4}$ or 168.75 or –5.

When you cut up a pie into sectors, fractions come in handy. At the dinner table you might ask, "Mom, after you cut Dad's piece—which is one-quarter of the pie —could I please have the rest?" Your mother, being good in mathematics, does the computation: $1 - \frac{1}{4} = \frac{3}{4}$ and hands you three-quarters of the pie.

But there are times when decimals are more useful than fractions. For example, the bike cost Fred $1,935.06. You could write that as $1935 $\frac{6}{100}$ but that looks a lot messier.

Can you imagine what a car odometer* would look like if instead of displaying:

40528.0
40528.1
40528.2
40528.3

it displayed fractions like:

40528 $\frac{1}{8}$
40528 $\frac{1}{4}$
40528 $\frac{1}{3}$
40528 $\frac{1}{2}$
40528 $\frac{9}{16}$

* An odometer is the gauge that tells you how far you have gone.

In our number system the position of the digits makes a difference. Would you rather have $18 or $81? Both have the numerals *1* and *8*, but where the *1* and *8* sit makes a big difference.*

We use the **base ten** system. When we look at a number like

1,935.06

the digit to the left of the decimal (the *5* in this case) is 5 ones. As we move to the left, each digit is "worth" ten times as much. As we move to the right, each digit is "worth" one-tenth as much.

1,935.06 = 1 thousand
+ 9 hundred
+ 3 tens
+ 5 ones
+ 0 tenths
+ 6 hundredths.

We could say that the base ten system is "handy" because—well, look for yourself:

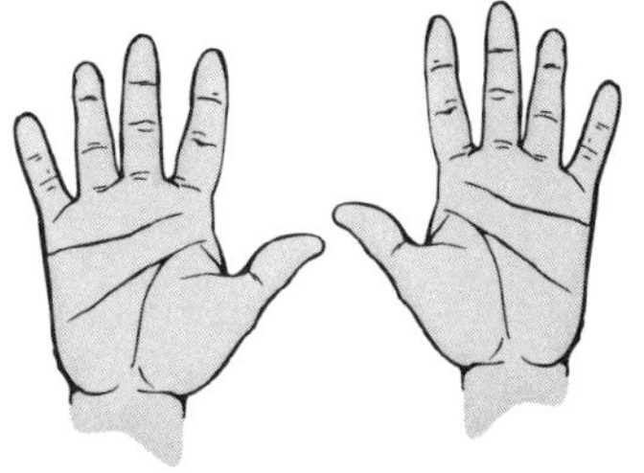

But other base systems have been used over the years. Many ancient cultures used a base 20 system** (fingers and toes). In the vigesimal system, when you wrote **35**, that meant 3 score + 5 ones. A **score** means 20. So **35** in the base 20 system is the same as 65 in our base ten system.

Traces of the vigesimal system remain in President Lincoln's famous words, "Four score and seven years ago. . . ."

It's time to take a little break. It's time for *Your Turn to Play*. I've been having all the fun so far. It's only fair that you get your chance.

The answers are listed on page 185, but please play with the questions a little bit (that is, *answer them in writing*) before you look at my answers.

---

* In fancy language, we call this a **place-value system** or, even fancier, a **positional numeration system**.

** More fancy language: vigesimal numeration system (vy-JESH-eh-mul).

Your Turn to Play

1. Write 87 in the vigesimal system.

2. Another really popular numeration system was the base 12 (duodecimal system). There are lots of places in everyday life that reflect the old base 12 system. Can you name three?

3. The oldest known place value system is the Babylonian sexagesimal system (base 60). Can you think of a couple of places in everyday life today that reflect that old system?

4. $4\frac{2}{3} - 2\frac{3}{4} = ?$

Complete solutions are on page 185

Please write out all your answers before you look at mine.

Chapter Two
Adding Decimals

One important decision Fred faced was how large to make his robot. One possibility was to make it one inch tall. Then he could carry it around in his pocket. If he put it on the top of his desk, he could watch it pick up pencils and carry them. He could teach it how to open envelopes.

Fred looked through the huge pile of parts on his office floor. Some of the pieces were much too large for a one-inch robot. Then he found a small bag. Perfect! he thought and opened the bag.

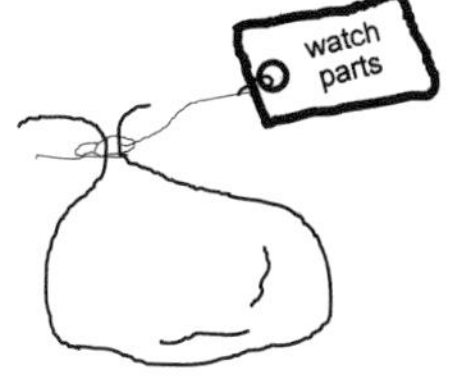

With a tiny pair of tweezers he carefully took out a little spring and put it on his electronic scale. It weighed 0.007 grams.

0.007

A gram is about the weight of a raisin.

$0.007 = 0 \text{ tenths} + 0 \text{ hundredths} + 7 \text{ thousandths} = \frac{7}{1000}$

He then weighed a tiny gear.

0.09

And a little watch motor.

13.3029

How much would all three items weigh? One way* to find out would be to put all three items on the scale. Another way would be to add up 0.007, 0.09, and 13.3029.

✶ *Weigh* and *way* are called homonyms (HOM-oh-nims). They sound alike. Some people make collections of homonyms. When they find three homonymic words, they are really happy. One triple is *to*, *too*, and *two*. Another is *merry*, *marry*, and *Mary*.

Adding decimals is easy to do. The trick is to line up the decimals.

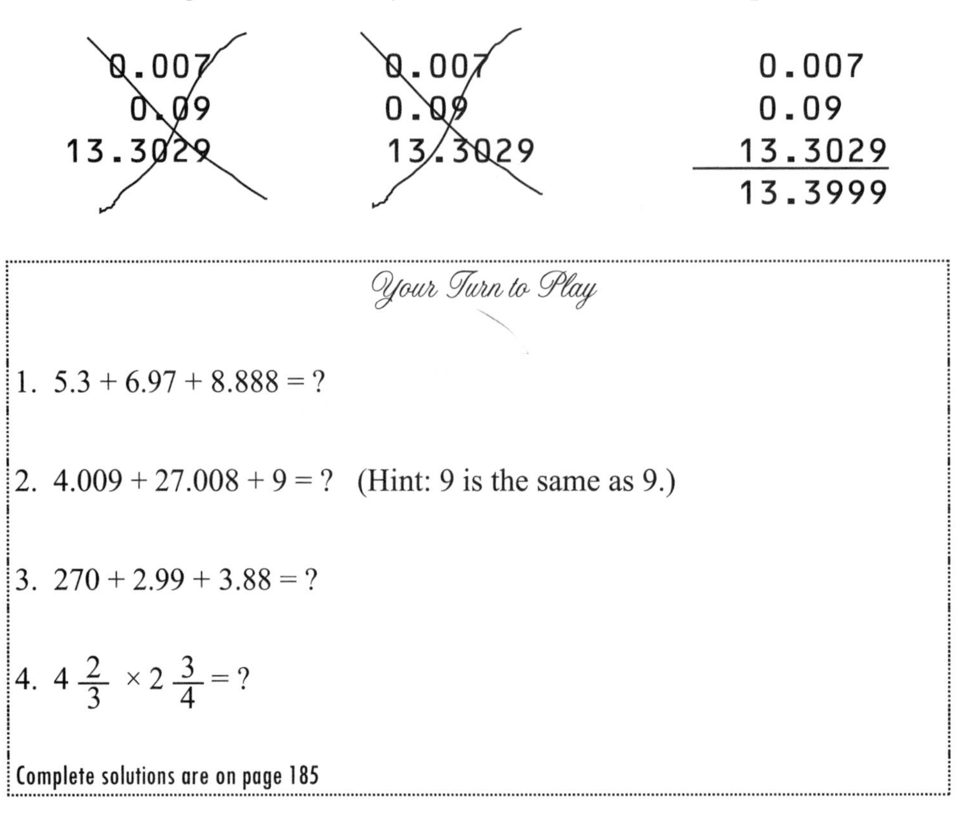

Your Turn to Play

1. $5.3 + 6.97 + 8.888 = ?$

2. $4.009 + 27.008 + 9 = ?$ (Hint: 9 is the same as 9.)

3. $270 + 2.99 + 3.88 = ?$

4. $4\frac{2}{3} \times 2\frac{3}{4} = ?$

Complete solutions are on page 185

Chapter Three
Subtracting Decimals

Fred's eyes are really good at looking at very tiny things. Just look at those dots! At the age of five-and-one-half, presbyopia* isn't a problem.

He put his eye right up to that tiny gear—the one that weighed 0.09 grams. He saw something. It was crawling. It was a little insect.

Image of tiny gear
multiplied 1000 times

Fred took his tweezers and put the little bug on a scale. It weighed 0.00072 grams. That meant that the gear really didn't weigh 0.09 grams, but instead it weighed 0.09 – 0.00072 grams.

Subtraction of decimals is just like addition. You line up the decimals:

$$\begin{array}{r} 0.09 \\ -\ \underline{0.00072} \end{array}$$

Oops. What do you subtract the 072 from? You need a trick, and I just happen to have one on the next page.

---

* Presbyopia (press-bee-OH-pea-eh) is often one of the signs of aging. (You do hope to get old, don't you?) The lenses in your eyes get less flexible. Very young children can flex their lenses and focus on very near objects and on distant objects. Older people often need reading glasses or bifocals to be able to focus at various distances.

The Trick of Adding Zeros

Normally you can't just go around adding zeros to a number. If you put a zero on 81 you get 810, which is a lot larger.*

But you can add zeros to the digits on the right of the decimal. Look at 0.09. It means 0 tenths + 9 hundredths.
Look at 0.09000. It means 0 tenths + 9 hundredths + 0 thousandths + 0 ten-thousandths + 0 hundred-thousandths.

0.09 means the same as **0.09000**

You can always add zeros on the *right* of the last digit that is on the *right* of the decimal. (You could call this the Double-Right Rule. Nobody else calls it that—but you could call it that.)

So 9878.3 = 9878.300
0.003 = 0.00300000000000000000000000
8.0 = 8.00
72 = 72.000 (because 72 = 72.)

And now we can subtract.

$$\begin{array}{r} 0.09 \\ -\ 0.00072 \\ \hline \end{array} \quad \Rightarrow \quad \begin{array}{r} 0.09000 \\ -\ 0.00072 \\ \hline 0.08928 \end{array}$$

So the gear alone weighs 0.08928 grams.

---

* ". . . a lot larger." That's the correct spelling. Some people want to make "a lot" into a four-letter word *alot*, but there is no such word. If you write *alot,* it means that you never read *Life of Fred: Decimals and Percents*.

Your Turn to Play

1. And, of course, you can always get rid of those extra zeros that are on the right of digits that are on the right of the decimal point. (Another Double-Right Rule?)

   79.05000 = ?

2. 3.07 − 1.008 = ?

3. A question for English majors: Suppose you wanted to say that the digit 9 followed by a decimal point is the same as a plain 9 without a decimal point.

   You write something like *9. is the same as 9.* To my eye, that seems a bit strange.

   Or, once, when I was in high school, I wrote in an essay the sentence: *Rocky owed Sylvia $2..* The first dot was for the decimal, and the second dot was the period. The teacher marked it wrong.

   Okay, English majors, here is your multiple-choice question: Is English harder than math? Here are your choices: ☐ yes or ☐ yes.

4. $4\frac{2}{3} \div 2\frac{3}{4} = ?$

Complete solutions are on page 186

The little tiny bug that Fred had put on his electronic scale started to crawl away. *It's so cute* Fred thought. *Maybe I can make it into a pet.* With his tweezers he transferred it to his microscope to get a better look at it.

"What shall I call you, little bug?" Fred asked as he adjusted his microscope. "You have to have a name, if you're going to be my pet. Maybe I'll call you Billy. Yes. Billy Bug. That will be your name."

When Fred looked into the microscope, he couldn't see Billy. He had crawled off the microscope.

Fred found him and put him back under the microscope. A little drop of glue kept Billy from wandering away again.

Everyone knows that microscopes don't make things weigh more. They just make things look larger. Billy still weighed 0.00072 grams. But under Fred's microscope, he looked 100 times larger.

He used to be 0.0385 cm long.* Under the microscope Billy looked like he was 3.85 cm long. (To multiply by 100, you move the decimal two places to the right.) Another way of saying that is: If you move the decimal two places to the right, the number will be a hundred times larger.

Suppose you wanted to make 5 a hundred times larger. There's no decimal to move, so the first step is to put one in. Then using the Double-Right Rule, stick in some zeros. Now you can move the decimal.

5 → 5. → 5.00 → 500.

✶ The abbreviation for centimeter is cm. A centimeter is a little less than a half-inch long. It is about $\frac{2}{5}$ of an inch. More precisely, one centimeter is approximately 0.3937 inches.

You see the line segment right above—the one that divides the text from this footnote. That segment is very close to 5 cm long.

Your Turn to Play

1. Multiply each of these by a hundred.

   0.235 88.8 0.01 71

2. Multiply each of these by a thousand.

   5 0.33 a million

3. *Multiplying by a thousand* is a fixed rule. It is a **function**. What is the opposite function (known as the **inverse function**)?

4. What is the inverse function of *divide by ten*?

5. A question for English majors: The Double-Right Rule was *You can add zeros on the right of the last digit that is on the right of the decimal.* When I want to divide a number like 0.6 by a thousand, I'm going to need the Double-Left Rule. State that rule.

SMALL DISCUSSION: When you were in first grade, the only questions they would ask you were the ones where they had already given you the answer. They would tell you that 7 × 8 equals 56, and then they would ask you what 7 times 8 equals.

You are no longer a baby. You have never been told what the Double-Left Rule is. You are being asked to figure it out. "Tape recorder students" in school are those kids who just memorize everything the teacher says (like good little tape recorders) and repeat it back on the tests.

6. Divide each of these by a thousand.

   7777 0.02 0.

7. $0.4 - 0.004 = ?$

8. $6\frac{2}{5} + 2\frac{1}{2} = ?$

9. If you wanted to multiply some number by a million, how many places to the right would you move the decimal?

10. If you wanted to divide some number by a billion, how many places to the left would you move the decimal?

11. Express an angle of 7° in minutes.

Complete solutions are on page 186

Chapter Five
Pi

Billy Bug stayed right where Fred had glued him. Fred looked into the microscope to see his new pet. One really important part of having a pet Fred thought was to be able to see him. Fred set the microscope on low magnification. This is what he could see:

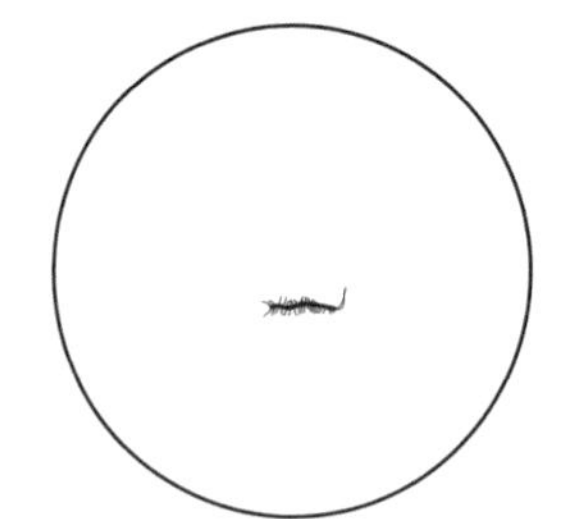

The low power setting on the microscope wasn't working very well. All he could see was that Billy's feet were moving. Maybe Billy is trying to wave to me.

Fred felt bad about having glued his pet in place. Maybe I could build a little fence around him. Then I could unglue him, and he could run around and play. That would be nice.

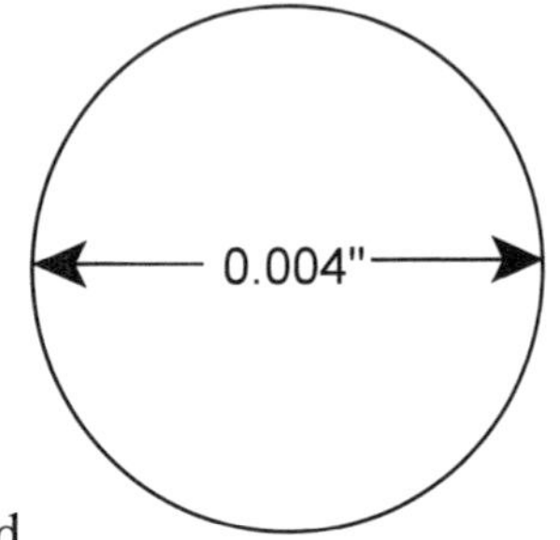

Fred measured the diameter of the circle. He needed to know the circumference* of the circle so he could build a fence. He was going to use a piece of string for the fence, and he needed to know how long to cut it.

In *Life of Fred: Fractions*, we took the diameter and multiplied it by $3\frac{1}{7}$ in order to find the circumference. The formula was $C = (3\frac{1}{7}) \times d$, where C is the circumference, and d is the diameter.**

But how do you multiply $3\frac{1}{7}$ by 0.004? That's a mess.

---

* Circumference is the distance around the circle.

** Using letters like *C* and *d* is the start of algebra. Because the three pre-algebra books beginning with *Life of Fred: Pre-Algebra 0 with Physics* are next in the series, we are getting you warmed up for algebra in advance.

Instead of $3\frac{1}{7}$ we can use 3.14, because they are approximately equal to each other.

In math you know the symbol for equals. $2 + 2 = 4$. There is also a symbol for *approximately equal to.* It is a wavy-looking equals sign.

$$1.99 + 1.99 \approx 4$$

So $3\frac{1}{7} \approx 3.14$, and we can write $C = 3.14 \times d$.*

Actually, the circumference is only approximately equal to $3\frac{1}{7}$ times the diameter. $C \approx (3\frac{1}{7})d$.

The circumference is only approximately equal to 3.14 times the diameter. $C \approx 3.14d$

Either of these work fine for most everyday uses.

Wait! Stop! I am the reader of this book, and I want to talk back to you.

Go ahead. What's on your mind?

I have been quietly reading your book now for about four chapters, but there is something I need to ask.

Okay. Ask it. Billy Bug isn't going anywhere right now. This is a perfect time to take a break.

You have been talking about $3\frac{1}{7}$ ***and*** 3.14 ***being just approximately the right number to multiply the diameter by in order to get the circumference. What if I want a better answer?***

That's easy. Use $C = 3.1416d$ instead.

Finally! Why didn't you just say that in the beginning, instead of using the $C \approx 3.14d$ ***stuff?***

Oops. I think I misled you. I should have written $C \approx 3.1416d$. You asked for a better answer, and 3.1416 is better than 3.14.

✶ In algebra, we stop using "×" for multiplication. It gets too confusing if we write x×y. The "x" gets mixed up with the "×".

Two letters next to each other means multiply. xy means x×y.

A number and a letter next to each other means multiply. 3.14d means $3.14 \times d$.

Better? I'm tired of "better." I want the right answer, the perfect answer, the exact answer. Do you hear me?

Loud and clear. You asked for it. Here is the perfect, exact answer. *If you want to get the circumference of a circle, you multiply the diameter by π.* $C = \pi d$.

What's that "π" thing? It looks like a "t" with two feet.

π is a Greek letter. It's pronounced "pie" and written "pi." π is a number. $\pi \approx 3\frac{1}{7}$ and $\pi \approx 3.14$ and $\pi \approx 3.1416$.

Pi doesn't look like a number. It looks like—I don't know—it looks like something from Mars.

When you could only count up to 3, the numeral *9* looked really funny to you. We have several more funny-looking numbers in math. In advanced algebra, we will show you a number that is not positive or negative or zero. We'll call it "i".

Forget i. Let's get back to π for a moment. Do you have any better approximation for π than 3.1416?

Sure. But you wouldn't want to see it. Pi is an unending, non-repeating decimal. It goes on forever.

Hey! I just asked. Show me more of pi.

You asked. I answer. $\pi \approx 3.14159265358979323846264338327950288419716939937510582\ldots$

Ha! I caught you in a lie! You're just making up those digits.

No I'm not. I copied them out of a reference book. Those are the real digits of π.

Look. A minute ago you said $\pi \approx 3.141\underline{6}$. ***Then when I asked you to show me more digits, you wrote*** $\pi \approx 3.141\underline{5}9\ldots$ ***The fourth decimal place was*** 6 ***and then, suddenly, it was a*** 5.

It's called **rounding**. When you want π to four decimal places, the best answer is $\pi \approx 3.1416$. If you want five decimal places, it's $\pi \approx 3.14159$.

If you want to get rid of the 9 in $3.1415\not{9}$
then the best possible four decimal places are 3.1416

It's called **rounding up**. That's because 3.14159 is closer to 3.1416 than it is to 3.1415.

The rule for rounding is: *If the digit you are getting rid of is 5, 6, 7, 8, or 9, then round up. If the digit you are getting rid of is 0, 1, 2, 3, or 4, then just get rid of it.*

$\doteq$ means *equals after rounding off.*

Examples:

Round 88.278 to 2 decimal places. 88.27~~8~~ $\doteq$ 88.28

Round 6.481 to 2 decimal places. 6.48~~1~~ $\doteq$ 6.48

Round 0.394396 to 3 decimal places. 0.394~~3~~96 $\doteq$ 0.394

Round 385 to the tens place. 38~~5~~. $\doteq$ 390.

Round 1,683,557 to the nearest million. 1,~~6~~83,557 $\doteq$ 2,000,000

Your Turn to Play

1. Round π to the nearest thousandths place.

2. Round each of these to the nearest tenth.

   0.8888 3.90 0.02 π 776.

3. $6\frac{2}{5} - 2\frac{1}{2} = ?$

Complete solutions are on page 187

Those of you who have read *Life of Fred: Fractions* know exactly what this bridge means. You can skip the rest of this page. Just turn to the next page and cross that first bridge.

The rest of you deserve a little explanation.

You bet! I just finished the Your Turn to Play in Chapter 5, and I was expecting Chapter 6. Where's Chapter 6?

We are at **The Bridge**.

After every five chapters, we give you the chance to show that you haven't forgotten what you've learned. After you prove that you remember, then you may continue to the next chapter.

There are ten questions on **The Bridge**. Answer at least nine of them correctly, and you can move on to Chapter 6.

Nine!

Yes. At least nine-tenths of the questions right. (Later on in this book, we'll call that 90%.) You are proving that you have mastery (not mystery) of the material.

Um. I hate to mention this, but what if I happen to get only eight right? Or seven? Or six?

If you don't pass on the first try, then correct your errors on that bridge. That will give you the right to have a second try at crossing the bridge. I've written another ten questions, which I call the "second try," and you can have another shot at showing you have mastery of the math.

And if . . .

You don't even have to ask. I wrote a third set of ten questions. And a fourth set of ten questions. And a fifth.

You will get lots of chances to show that you have got what it takes.

Okay. I'm ready. Bring it on. Where's this bridge thing?

Just turn the page.

The Bridge

from Chapters 1–5 to Chapter 6

first try

> Goal: Get 9 or more right and you cross the bridge. Please do all the problems before looking at my answers. It's okay to look back to previous pages when doing a bridge.

1. We use the base ten system. The vigesimal system is what base?
2. $87 + 22.09 + 0.77 = ?$
3. A box of rice weighed 453.03 grams. I took out one grain of rice from the box. The grain weighed 0.005 grams. Now how much does the box weigh?
4. Billy Bug now weighs 0.00072 grams. When he was born he weighed one hundred times less. How much did Billy weigh when he was born?
5. Express 44° in minutes.
6. What is the inverse function of *divide by a million*?
7. If the diameter of a circle is ten feet, which of the following would be its circumference: $10 + \pi$ or $10 - \pi$ or 10π or $\frac{10}{\pi}$?
8. Round 1339 to the tens place.
9. The gear and Billy Bug together weigh 0.09 grams. Round that to the nearest tenth of a gram.
10. Billy had a dream. (Bugs can dream, can't they?) He dreamed that he was being chased by a monster bug who was a thousand times heavier than he. Billy weighs 0.00072 grams. How much does the monster bug weigh?

The monster bug in Billy's Dream

(The answers to **The Bridge** are given in the back of this book beginning on page 172.)

The Bridge

from Chapters 1–5 to Chapter 6

second try

---

1. Divide 2.09 by a thousand.

2. What is the inverse function of *add six*?

3. 0.032 + 5 + 3.07 = ?

4. Write 62 in the vigesimal system.

5. Round 8.956 to 2 decimal places.

6. I had a bunch of grapes that weighed 200.17 grams. I ate one of the grapes that weighed 8.003 grams. Now how much do the remaining grapes weigh?

7. When a grape dries up and becomes a raisin, it weighs one-tenth of its original weight. Suppose a grape weighed 10.07 grams. How much will it weigh when it turns into a raisin?

8. If a grape weighs 10.07 grams, round that to the nearest tenth of a gram.

9. An ordinary grape costs 0.3¢. A premium grape costs ten times as much. How much does a premium grape cost?

10. Suppose a grape weighed 10.07 grams. How much would a million of those grapes weigh?

The Bridge

from Chapters 1–5 to Chapter 6

third try

1. 429 + 0.398 + 8.254 = ?

2. Divide 88 by a million.

3. What is the inverse function of *losing 4 degrees of temperature*?

4. A bowl of regular ice cream costs $1.07 and a bowl of premium ice cream costs ten times as much. How much does a bowl of premium ice cream cost?

5. My temperature was 98.6°. After I had some ice cream, it dropped by 0.07°. What was my new temperature?

6. I weigh one hundred times as much as a bowl of ice cream. The bowl of ice cream weighed 992.3 grams. How much do I weigh?

7. One hundred and twenty minutes of angle is equal to how many degrees?

8. If a bowl of ice cream weighs 992.3 grams, how much is that to the nearest gram?

9. If a bowl of ice cream weighs 992.3 grams, and Billy Bug ate one-millionth of that, how much did he eat?

10. Round 4.4399 to 2 decimal places.

The Bridge

from Chapters 1–5 to Chapter 6

fourth try

1. What is the inverse function of *multiply by 9.4*?

2. 37 – 0.03 = ?

3. Round 856.7724 to the tens place.

4. Divide 93.2 by a billion.

5. Round π to the nearest hundredth. (π ≈ 3.1415926535)

6. Convert 360 seconds of angle into minutes of angle.

7. If a pitcher can throw a ball 72.05 mph (miles per hour), how fast is that rounded to the nearest tenth?

8. 0.03 + 3 + 0.387 = ?

9. If the ball weighs one-hundredth as much as the pitcher, and the pitcher weighs 98 lbs.*, how much does the ball weigh?

10. If the pitcher weighs 98 lbs. at the beginning of the game and loses 1.2 lbs. during the game, how much does she weigh at the end of the game?

✶ lb. is the abbreviation for pound.

The Bridge

from Chapters 1–5 to Chapter 6

fifth try

1. $7.92 + 8 + 0.393 = ?$

2. Divide 0.439 by a million.

3. Round 873.00389 to 4 decimal places.

4. Express 5° in minutes.

5. I planted a lot of daisies in my front yard right near the lawn. Their average height was 12.3 inches. I accidentally ran over them with my lawnmower. Their average height is now 2.08 inches. How much height did they lose?

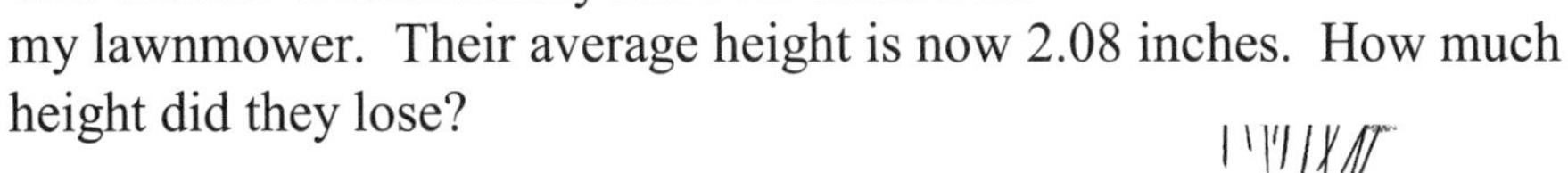

After being mowed

6. $50 - 3.98 = ?$

7. Originally, those daisies took 43 minutes to plant. It took one-thousandth of that time to mow them down. How long did it take to make this mistake?

8. I went to the store to buy some plastic flowers to replace the daisies that I mowed down. One hundred plastic daisies cost $29. How much does one plastic daisy cost?

9. My friends suggested that maybe I needed new glasses because I was mowing down the daisies. I went to the optometrist. My old glasses had a correction of 2.2 diopters. The new glasses have a correction of 4 diopters. What was the increase in diopters?

10. Round 3,539,982.007 to the nearest million.

Chapter Six
Multiplying Decimals

It was time to build the string fence around Billy Bug so that he wouldn't have to stay glued to the microscope. Fred knew that the diameter was 0.004 inches, and he knew C = πd, where C is the circumference and d is the diameter of the circle.

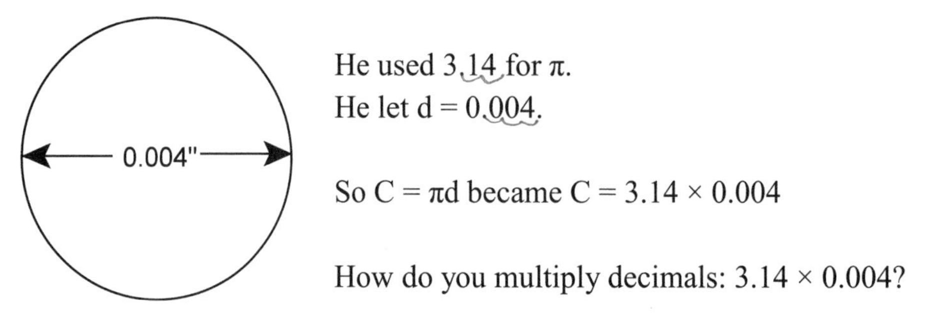

He used 3.14 for π.
He let d = 0.004.

So C = πd became C = 3.14 × 0.004

How do you multiply decimals: 3.14 × 0.004?

Here's where we *use* things we learned in the last couple of chapters: ① inverse functions, ② multiplying by ten, and ③ dividing by ten.

We know how to multiply whole numbers* together. The trick will be to turn 3.14 and 0.004 into whole numbers.

How do you change 3.14 into a whole number? You multiply by 100. That moves the decimal two places to the right.

How do you change 0.004 into a whole number? You multiply by 1000. That moves the decimal three places to the right.

Now our work is easy.

$$\begin{array}{r} 314. \\ \times \underline{\quad 4.} \\ 1256. \end{array}$$

What's the opposite of (or the inverse function of) *moving the decimal two places to the right* and *moving the decimal three places to the right*? You guessed it: *moving the decimal two places to the left* and *moving the decimal three more places to the left.* Take the answer of 1256

✶ The **whole numbers** = {0, 1, 2, 3, 4, 5, . . . }.

and move the decimal five places to the left.   1256. → 0.01256 That's the answer to 3.14 × 0.004.

$$\begin{array}{r} 3.14 \\ \times\ \underline{0.004} \end{array} \rightarrow \begin{array}{r} 314. \\ \times\ \underline{\quad 4.} \\ 1256. \end{array} \rightarrow 0.01256$$

Move 5 places to the right.   *Move 5 places to the left.*

Fred needed C = 3.14 × 0.004 = 0.01256 inches of string to build his circular fence around Billy. Perfect! thought Fred. He ran and got the string and cut it.

After building the string fence around Billy, Fred used his tweezers to free him from the glue. Billy, you must be so happy now. You are such a wonderful pet. I've got to find some food to feed you, and some toys for you to play with. And a little bed to sleep in. And I can read you a bedtime story if you like.

Fred wanted to get a better look at his pet, so he turned up the magnification on his microscope.

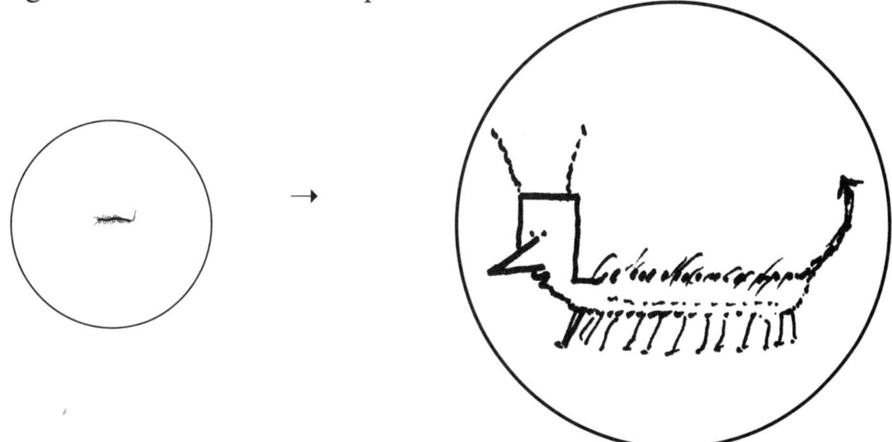

He's so cute! Oh look. His front two legs are still a little glued together. Fred grabbed a washcloth and ran down the hallway from his office. When he got to the restroom, he ran the water until it was warm—not too hot and not too cold—and wet the washcloth. He wrung out the cloth so that it was not too wet and not too dry. He wanted it just perfect so that he could wash the glue off Billy's front legs. As he ran back to his office, he called out, "Billy! I'm coming. I'll be right there."

When he looked into the microscope, he saw . . . nothing. Billy was gone. He had crawled over the string fence and had wandered off again.

nothing

Fred looked everywhere in his office. His beloved Billy was gone. Fred felt very lonely. He grabbed his blanket and one of the books off the shelf and headed to his "sad corner." The book was Prof. Eldwood's *When Bad Things Happen to Good Bugs*, 1848.

Luckily, Fred never noticed what was on the bottom of his shoe.

Your Turn to Play

1. $7.8 \times 0.007 = ?$

2. $6.59 \times 7.1 = ?$

3. $0.0003 \times 0.0003 = ?$

4. $0.0003 + 0.0003 = ?$

5. $0.0003 - 0.0003 = ?$

6. $6\frac{2}{5} \times 2\frac{1}{2} = ?$

7. Round 74.52222 to the nearest whole number.

Complete solutions are on page 187

Chapter Seven
Functions

After Fred had finished reading his book, he did a little crying and felt much better. He put his blanket back in a desk drawer and put the book back on the shelf. I'm ready for action now Fred thought. It's time I build my little robot.

He put the little bag marked "watch parts" and his pair of tweezers on his desk and sat down to work. This time there were no little insects on the gears to distract him. He took a little box out of the bag. This will make a perfect body. He attached a thimble for a head. He put in some interior plumbing, and it was starting to look like a one-inch tall robot.

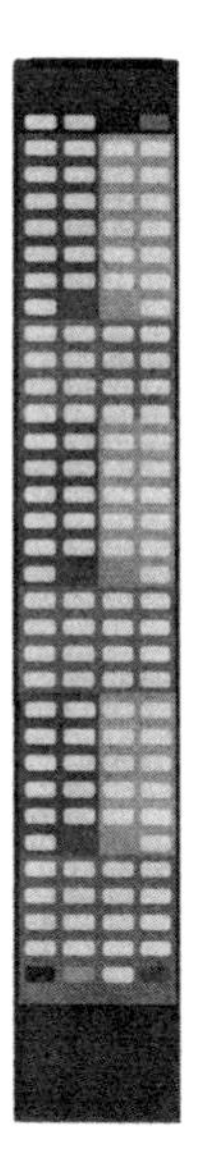

It was time to test his robot. Fred pressed buttons on his giant remote control. Nothing happened. When he pressed the button marked `raise arm`, the robot just stood there. When he pressed `start walking`, nothing happened.

Then he remembered what the scarecrow sang in the *Wizard of Oz* movie: ♫ If I only had a brain. ♪

How very silly of me Fred thought. He opened up the back of one of the computers he had on his desk and pulled out a chip.* Inserting the chip into the thimble head made a difference. Its eyes lit up. It raised its hand. It started walking.

Fred pressed the `stop walking` button. It stopped. He pressed `put down your hand`, and it put down its hand.

A function! Inverses! It was all right there.

✶ Generally, this is not recommended. Some computers don't seem to work as well when you pull out their chips.

Your Turn to Play

1. A **function** is any fixed rule or procedure. The important thing is that the rule doesn't change. If the robot is just standing there and you press the `start walking` button, it starts walking. It wouldn't be a function if sometimes it started walking and other times it started tap dancing.

Add 3 is a function. Every time you use that function on 5 you get 8. You don't get 8 some days and 9 on other days.

Subtract 3 is the inverse function of *add 3*. If you first do *add 3* and then you do *subtract 3*, you come right back to where you started.

What is the inverse function of `start walking`?

2. Is this a function? *Given a phonebook, select a name at random*.

3. Is this a function? *Square the number*. (Squaring a number means multiplying it by itself. The square of 7 is 49.)

4. Is *round the number to the nearest hundredth* a function?

5. [Hard question] Does *round the number to the nearest hundredth* have an inverse?

6. $6\frac{2}{5} \div 2\frac{1}{2} = ?$

7. If you are given two big fractions such as $\frac{391}{3907955077}$ and $\frac{224381}{708983}$ would it be easier to add them or to multiply them?

8. If the diameter of a circle is 8 feet, what is its circumference? (Use 3.14 for π.)

9. If the diameter of a circle is 8 feet, what is its radius? (The radius is the length from the center of the circle to the edge.)

10. $7.7 - 6.09 = ?$

Complete solutions are on page 188

Chapter Eight
Subtracting Mixed Units

Someone knocked on Fred's office door. Normally, Fred would just have said "Come in." But now, all he needed to do was push the `answer the door` button on the remote control.

The little robot walked across the room to the door. It reached up, but it couldn't touch the doorknob. It tried to pull the door open, but couldn't. With a helpless look on its face, it turned around and looked at Fred. Fred was silent. He wanted to see how smart his robot was.

The robot turned toward the door and said, "Please come in."

It was Betty. She is a twenty-year-old student at KITTENS University. Fred and Betty have known each other for years—ever since Fred first came to the university to be a math teacher.✶

"I was worried about you," Betty said as she came into the room.

The little robot ran and hid under Fred's desk. It didn't want to get stepped on.

"Did you get anything to eat today?" Betty asked. "When you and I and Alexander went out for lunch, you didn't sit down and eat with us. Instead, you got a job working in the pizza kitchen."

Fred thought for a moment. He had never really paid attention to what he ate or whether he was hungry. He said to Betty, "I had a little bite of a marshmallow this afternoon before I went through the car wash." (The adventures surrounding lunch, the job, the marshmallow, and the car wash all happened in *Life of Fred: Fractions*.)

✶ The story of how Fred became a math teacher at KITTENS University is all told in *Life of Fred: Calculus Expanded Edition.* After you finish *Decimals & Percents,* the next books are

Life of Fred: Pre-Algebra 0 with Physics (40 daily lessons)
Life of Fred: Pre-Algebra 1 with Biology (46 daily lessons)
Life of Fred: Pre-Algebra 2 with Economics (34 daily lessons)
Life of Fred: Beginning Algebra Expanded Edition (104 daily lessons)
Life of Fred: Advanced Algebra Expanded Edition (105 daily lessons)
Life of Fred: Geometry Expanded Edition (2 semesters)
Life of Fred: Trigonometry Expanded Edition (94 daily lessons)

and then you will be at *Life of Fred: Calculus Expanded Edition.* (4 semesters)
After these books, you will be at the equivalent of the fifteenth grade in math. You will have covered all of the twelve grades and will be in your third year of college.

Because of his poor eating habits, Fred had never grown much. At five-and-a-half years, he was only thirty-six inches tall. He weighed thirty-seven pounds. Most of the time, Fred would eat from the vending machines that were in the hall outside his office. A diet of doughnuts, candy bars, and Sluice soft drink doesn't offer much protein. Fred hadn't grown as fast as other kids.

"A bite of marshmallow! I thought so," Betty said. "Alexander will be by in a couple of minutes, and we're going to take you out to a place you've never been. A steak house. It's brand new. It just opened up. It's on Newton Way."*

Betty looked at Fred's office. Usually, it was very neat—books on the shelves, Fred's desk in the middle of the room, and a little sleeping bag rolled up under his desk. Today, however, it was a mess. There were wires, and gears, and motors, and electric plugs scattered all over his desk and floor.

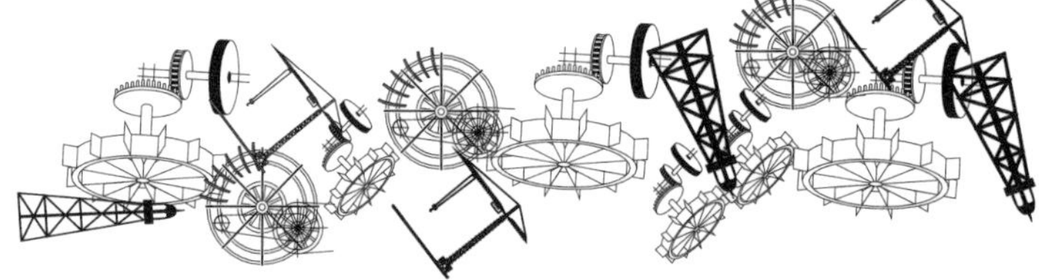

"Please excuse the mess. I have been busy building a little robot," Fred explained. He pressed `clean up the mess` on his remote control. The robot came out from under the desk and started pushing the junk into a corner of the room. Betty picked it up.

"It's so cute!" said Betty. "What's its name?"

Fred hadn't had time to name it yet. He thought quickly and said, "I call him Robert Robot."

Betty felt something wiggle in her hand. With hands on hips, the robot shouted, "Hey! I'm not a boy robot. I'm a girl. And my name isn't Robert. It's Roberta."

---

* Isaac Newton was one of the two inventors of calculus. Newton was, perhaps, the second best mathematician who ever lived.

Betty smiled. "That's still alliterative.✶ I like that. Roberta Robot. You are so cute."

The robot continued, "But my friends called me Bobbie Bot. Everyone knows that *bot* is short for *robot*."

Betty put Bobbie on the floor, and she continued pushing junk into the corner of the room.

Another knock on the door. Fred pushed `answer the door`. Bobbie shouted, "Please come in."

In strode✶✶ Alexander. "Who said that?" he asked.

"I did," came the answer from the corner of the room.

Bobbie's thimble head was five feet and eleven inches below the top of Alexander's head.

Your Turn to Play

1. Do you know how to subtract 1" (Bobbie's height of one inch) from 6' (Alexander's height of six feet)? You do it by borrowing.

		becomes		
	6 feet			5 feet 12 inches
−	1 inch		−	1 inch
				5 feet 11 inches

How much taller is Fred than Bobbie? (Fred is 3'.)

2. The doorway to Fred's office is 6' 8". Betty is 5' 9". What's the tallest hat that Betty could wear and just touch the top of the doorway?

3. Fred's office is almost six yards long. In fact, it is one inch less than six yards long. How long is it?

4. Suppose you died one second before you became a century old. How long would you have lived? (You may ignore leap years.)

Complete solutions are on page 189

✶ Alliteration means words starting with the same sound. For example, Fred's friendly face.

✶✶ *Strode* is the past tense of stride. To stride is to walk with long steps. Because Alexander is six feet tall, he doesn't just walk. He strides. Words that indicate action are called **verbs**. So *stride, sing, speak,* and *walk* are verbs. Their **past tenses** are *strode, sang, spoke,* and *walked.*

Chapter Nine
Sets

Alexander knelt down next to Bobbie and watched her work. She had pushed the gears into one pile, the wires into a second pile, and the motors into a third pile.

"Where did you get that?" Alexander asked Fred.

"I built it," Fred answered. "It—I mean she—even has a remote control."

"She?" asked Alexander.

"You can call me Bobbie," she said. "I'm currently cleaning up the room because Fred pressed the `clean up the mess` button on the remote control."

Alexander picked up the remote control. "Wow. This is heavy. It must weigh eight pounds. I've never seen one this large before. May I?" Alexander asked Fred, indicating that he would like to play with the remote control.

"Sure," said Fred.

Before Alexander could push any of the buttons, Bobbie came over to Alexander and said, "Excuse me." She reached under his knees and lifted him up! All 180 pounds of Alexander. Completely off the ground. Then she reached underneath and grabbed a tiny gear that he had been kneeling on. She set him down and tossed the gear on the gear pile.

Everyone in the room was silent.

Real silent.

All except Bobbie. She said, "All done. I've got the mess all cleaned up."

Betty finally spoke, "Could . . . you . . . please explain what you just did?"

Now you and I know that Betty was asking Bobbie how she was able to lift Alexander off the ground. Betty wanted to know how Bobbie was so strong. That's because you, I, and Betty are human beings. We don't expect one-inch tall robots to go around lifting really heavy things. We see something surprising, and we want to know why.

But look at it from Bobbie's point of view. Everything is equally surprising because she was only created twenty minutes ago. Betty had forgotten the Forty-sixth Rule of Robotics:

Robots can be terribly literal-minded.

Bobbie had just said, ". . . I've got the mess all cleaned up," and Betty asked, "Could you please explain what you just did." Bobbie must have thought to herself How dumb humans are. What is there to explain about cleaning up this mess? It was just a pile of gears, pile of wires, and pile of motors?

What do they want me to talk about?

Do they want me to tell them what a gear is? Probably not.

Do they want me to explain how to clean up? That's too easy. To clean up all you do is shove the gears into one pile, the wires into a second group, and the motors into a third collection.

That makes three sets of things.

I know! They want me to talk about SETS. Okay. I'll try to explain it to these people.

Here is what Bobbie Bot told them: (This type face is her lecture on the topic of sets.)

Look. This is a pile of gears. That makes a set. A **set** is any collection of anything. Look out the window. You see that flock of pigeons? That's a set.

The whole numbers is a set. W = {0, 1, 2, 3, . . . }

We say that "2 is an **element** of W," or "2 is a **member** of W."

"{" is a left brace. "}" is a right brace. You can describe a set by putting its members in between braces.

Fred, Betty, and Alexander were amazed. They were being lectured by a one-inch-tall robot. Fred and Betty sat down next to Alexander on the floor. Bobbie waited for them to get settled and then she continued:

A set can be any collection. Here is a set. { , , ♣ } If I call this set *A,* then I can say that ♣ **belongs to** *A*.

In symbols, this is written ♣ $\in A$.

The symbol "$\in$" means *belongs to*, *is an element of*, or *is a member of*.

And the symbol $\notin$ means *does not belong to*, *is not an element of*, or *is not a member of*.

Betty interrupted, "This is just like the equals sign. In arithmetic we can write $2 + 2 = 4$, and we can write $2 + 2 \neq 5$.

"And in the same way, I can write Fred $\notin$ { , , ♣ }."

Bobbie continued. Sometimes it is hard to list all the elements inside of braces. If there are only three elements in a set, it is easy to list them all. For example, { , , ♣ }.

But what if I wanted to list all the books in the San Francisco public library? If I tried to list them all inside of braces, I would be in trouble.

{*A.A. Milne : The Man Behind Winnie-The-Pooh*, Ann Thwaite, *A.A. Milne's Pooh Classics*, *A Novel*, Warhol, Andy, *A. Alekhine : Agony Of A Chess Genius*, Pablo Morán, *A B C : A Family Alphabet Book. . . .*}

There are too many books. I couldn't list them all.

Instead, I use **set-builder notation**.

I write { x | x is a book in the San Francisco public library}.

This means, "The set of all x such that x is a book in the San Francisco public library."

If I wanted to write down the set of all the monkeys in the zoo, I could write that set as { x | x is a monkey in the zoo}.

If I wanted to write down the set of all numbers greater than a million, I could write that set as { x | x > 1,000,000 }.

"{ x |" means "the set of all x such that. . . ."

Can anyone tell me when two sets are equal? When can we say that set A is equal to set B?

Betty answered, “When they have the same elements.”

That’s right. The order in which you list the elements of a set doesn’t matter. If A = { , , ♣ } and if B = { ♣, , }, then $A = B$.

Now if you’ll pardon me for a moment. . . .

At this point, Bobbie was starting to feel a little weak. After lifting Alexander off the ground and doing all this lecturing on set theory, her battery needed recharging. She headed over to an electrical outlet

and plugged herself in. The lights in the building dimmed for a moment. Bobbie smiled. For her, it was better than pizza.

Then she turned back to Fred, Betty, and Alexander to continue her lecture.

Wait! Stop! I, the reader of this book, have a question for you, Mr. Author.

Okay. Ask it.

I want to know how long this robot is going to lecture. There is lots of electricity in an outlet and Bobbie could go on lecturing about set theory for days.

Actually, she could go on for months. There’s lots to talk about. At some universities you can take a course in set theory that lasts for a whole year. What Bobbie covers in this Chapter 9 are some of the things you would cover in *the first day* of that year-long university course.

Bobbie continued:

Subset

Sometimes one set is included inside another set.
{1, 3, 5} is included in {1, 2, 3, 4, 5}.
Another way to say this is {1, 3, 5} is a subset of {1, 2, 3, 4, 5}.

{ 🍭 , 🚜 , ♣ } is a subset of { 🍭 , 🚜 , ♣, ✈ }.

{6, 8} is a subset of {6, 8}.

{ x | x is an even number } is a subset of { x | x is any number}.

{✄, ✉, ❁} is a subset of { ✉, ❁, ✄}. The order doesn't matter. What is important is that every element of the first set is an element of the second set.

Official Definition

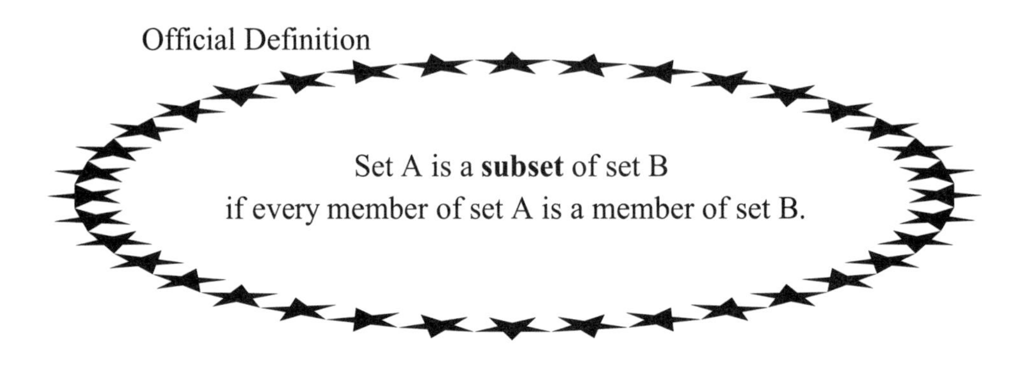

Set A is a **subset** of set B
if every member of set A is a member of set B.

The set of all roses is a subset of the set of all flowers.

{ x | x is chocolate} is included in { x | x is an ice cream flavor}.

Mathematicians like to save work. Instead of having to write, "Set A is a subset of set B," we abbreviate that as $A \subset B$. That saves about one second of writing.

For every set G, it is always true that $G \subset G$. In English that means that everything that is in G is in G. (You already knew that, but writing it in symbols, $G \subset G$, makes it look much more fancy.)

The set of all numbers that are less than 4 is a subset of all numbers that are less than 10. In symbols: $\{ x \mid x < 4\} \subset \{ x \mid x < 10\}$.

And now we come to my favorite set. It is the world's smallest set. In fact, it is soooooo small that it is a subset of every set. It's called the . . .

Empty Set

The empty set is the set that has no elements in it.
It looks like this: { }.

It's like a bag with nothing inside of it.

Since it takes so long to write braces, the empty set, { }, is sometimes abbreviated as ∅.

{ x | x is a person who is 50000 feet tall} is a set that contains no members. So { x | x is a person who is 50000 feet tall} = ∅.

"Set theory is such a big and important topic," Bobbie said. "I have been thinking of writing a whole book about it."

Fred commented, "That's quite a statement for something—I mean, someone—who was just created forty-five minutes ago."

"I've even thought of the title of my book and how the cover would look," continued Bobbie.

"That's kinda cute," Fred told her.

Alexander turned to Betty and said, "There are copyright laws. I hope Bobbie doesn't get sued."

Bobbie giggled. "I can't be sued. Only people and corporations can be sued. I'm a machine."

Alexander looked at his watch. "In a half hour, we have to get to the steak house. We have reservations at 5:30."

"No problem," said Bobbie. She started talking faster and sped* through the first chapter of her book.

There are three ways to combine sets. This is a lot easier than arithmetic where you had four ways to combine numbers: addition, subtraction, multiplication, and division. In arithmetic you had to learn your addition tables and your multiplication tables. $7 \times 8 = 56$. In set theory you don't have to memorize tables.

Union

{1, 2, 3} union {2, 4, 5} = {1, 2, 3, 4, 5} Combine together everything that was in at least one of the two sets. The spelling rule for listing sets: Don't list an element more than once. Don't write {1, 2, 2, 3, 4, 5}.

We abbreviate "union" as ∪.

{✂, ✉, ✽} ∪ {✝} = {✂, ✉, ✽, ✝}

Intersection

{1, 2, 3} intersection {2, 4, 5} = { 2 } Combine together everything that was in both of the sets. We abbreviate "intersection" as ∩.

{✂, ✉, ✽} ∩ {✝} = ∅

{ x | x is a letter in "Bobbie"} ∩ {a, c, e, f, g} = {e}

{ 🍬, 🚜, ♣ } ∩ { 🍬, 🚜, ♣ } = { 🍬, 🚜, ♣ }

✶ *Sped* is the past tense of the verb *speed*. *Speeded* is also the past tense. You get your choice. I chose *sped* because it's shorter. ***Wait! Stop! I, your reader, have a thought. It's true that sped is shorter than speeded, but you used two lines of footnote to explain why you used four letters instead of seven.*** Good point. Or maybe I should say good pt.

Difference

Set A – set B is defined as everything that is in A but not in B.
{1, 2, 3, 4} – {2, 6, 7} = {1, 3, 4}
{✄, ✉, ✱} – {✝} = {✄, ✉, ✱}

Alexander said, "It looks like it's time to go. Bobbie, is there anything we can get you at the steak house? I know you only eat electricity."

"Yes there is," answered Bobbie. "If you have salads, please bring me a little salad oil. My joints could use a little grease."

Your Turn to Play

1. Name an element of the set of whole numbers.

2. Suppose A = {Sam, Pat, Chris} and B = {Pat, Joe}.
   Find $A \cup B$, $A \cap B$, and $A - B$.

3. Suppose Bobbie weighed 6 lbs. 1 oz. and lost 2 oz. of her oil. How much would she weigh? (1 lb. = 16 oz.)

4. $9\frac{1}{2} + 5\frac{4}{5} = ?$

Complete solutions are on page 189

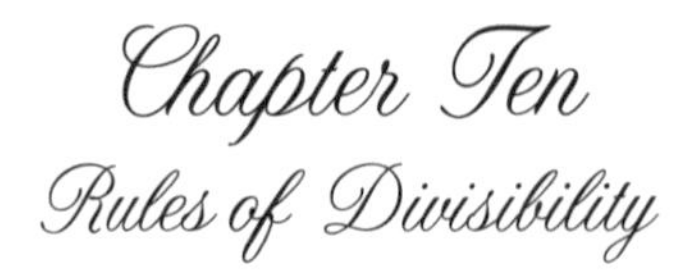

Alexander set the remote control on the floor [☞ important] and got up. Betty and Alexander headed out the door first. Fred said, "Nighty-night" to Bobbie, turned off the light [☞ important] and closed the door. They walked down the two flights of stairs of the Math Building and out into the night. They were off to the new steak house on Newton Way.

Alexander strode down the street. Betty walked. And Fred jogged to keep up with them. A gibbous* moon lit their way.

As they passed 165th Avenue, Betty remarked, "165 is divisible evenly by 5. But I guess you all know that. When you count by fives—5, 10, 15, 20, 25—all the numbers end in either 0 or 5."

> Rule: If the last digit is 0 or 5, it's divisible by 5.

As they passed 166th Avenue, Alexander said, "166 is divisible evenly by 2, because 166 is even. But I guess you all know that."

> Rule: If the last digit is 0, 2, 4, 6, or 8, it's divisible by 2.

As they passed 2817th Avenue, Fred said, "2817 is divisible evenly by 3. But I guess you all knew that."

"What!" exclaimed Alexander. "No I didn't know that. What's the trick?"

Fred answered, "If you add up the digits, and the sum is divisible by 3, then the original number is divisible by 3. So I added up the digits—2 + 8 + 1 + 7—and got 18. Because 18 is evenly divisible by 3, so is 2817."

> Rule: If the sum of the digits is divisible by 3, so is the number.

---

* A gibbous (GIB-us) moon is one that is more than half full.

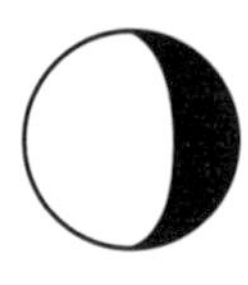

After walking a couple more blocks, Fred remarked, "I wish that they had spelled *divisibility* just a little differently."

"What do you mean?" asked Betty.

"Just spell the word out loud. It's got a great rhythm.
di vi si bi li and then they spoil it with ty. I wish it were spelled divisibiliti."

"You're silly, Fred," said Betty.

Your Turn to Play

1. Is this number divisible by 3? 10000000000000020000900000

2. If you have a number that is divisible by 3, and you multiply that number by a billion, will the answer be divisible by 3?

3. Name the smallest whole number that is evenly divisible by 2, 3, 5, 17, and 39763.

4. $9\frac{1}{2} - 5\frac{4}{5} = ?$

Complete solutions are on page 190

The Bridge

from Chapters 1–10 to Chapter 11

first try

Goal: Get 9 or more right and you cross the bridge.

1. If the diameter of a dinner plate is 7.4", what is its circumference? (Use 3.14 for π.)

2. At the dinner table you are given one of your Aunt Alice's famous raisin muffins.

Recipe
65.3 grams stone-ground flour
4.75 grams cheese powder
2 cups cement
1 raisin
50.3 milliliters water
Mix well and bake at 600° for 3 hours.

With a steak knife you manage to cut off a slice of her muffin. *Cutting off a slice of Aunt Alice's famous raisin muffin* is a function. Does it have an inverse?

3. There are 8 fluid ounces in a cup. If Aunt Alice had used 3 ounces less of the cement, how much would she have used? (See the above recipe.)

4. Let W = the set of whole numbers. We know that $5 \in W$. What is the symbol you would put in the blank:

An Aunt Alice raisin muffin __[blank]__ W.

5. Which of these numbers—2, 3, or 5—divides evenly into 780001?

6. How much do the stone-ground flour and cheese powder weigh together? (See the above recipe.)

7. How much cheese powder is needed to make 7 muffins? (The recipe makes one muffin.)

8. Round 329.0892 to the nearest hundredth.

9. Round 329.0892 to the nearest hundred.

10. $8.9 + 33 + 0.09 + 0.99 = ?$

The Bridge

from Chapters 1–10 to Chapter 11

second try

---

1. Let A = {Chris, Jackie, Sig} and B = {Chris, Tracy, Drew}. Find $A - B$.

2. Which of these numbers—2, 3, or 5—divide evenly into six billion?

3. $783.03 + 29 + 0.98 = ?$

4. $8.001 \times 0.007 = ?$

5. If Tracy weighs 45.2 kg and Drew weighs 38.8 kg, how much heavier is Tracy than Drew?

6. If the diameter of a pizza is 14.2", what is its circumference? (Use 3.14 for π.)

7. What is the inverse of the function *lift the pizza three inches upward*?

8. A pizza weighed 4 pounds before baking. During baking, it lost 9 ounces. How much did it weigh after it was baked? (1 lb. = 16 oz.)

9. Round 0.00352 to the nearest tenth.

10. $8888 + 0.8888 + 88.88 = ?$

The Bridge
from Chapters 1–10 to Chapter 11

third try

---

1. $55 \times 0.005 \times 0.7 = ?$

2. *At a gas station you fill up your car with gasoline* is an action. Does it have an inverse? If so, what is it?

3. Subtract 4 feet 9 inches from 44 feet 3 inches.

4. What is the smallest element in the set W? (W = the whole numbers.)

5. Is there a largest member of W? If so, name it. If not, explain why not.

6. $0.0872 + 0.993 + 4 = ?$

7. Your car can hold 22 gallons of gas. You have 0.04 gallons of gas in your car when you drive into the gas station. How much gas will it take to fill up your car?

8. If the diameter of a circle is 0.077", what is its circumference? (Use 3.14 for π.)

9. Divide 47 by a million.

10. Which of these numbers—2, 3, or 5—divide evenly into 1000000000345?

The Bridge

from Chapters 1–10 to Chapter 11

fourth try

---

1. Joe and Darlene are two other students whom Fred has known for years. Joe and Darlene went out fishing once, but Joe wasn't very good at rowing his boat. He rowed it in a circle that had a radius of 20.3 feet. What was the diameter of that circle?

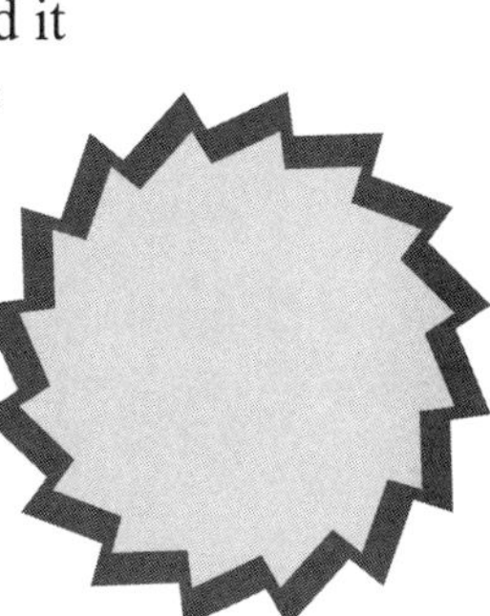

2. Continuing problem 1, what was the circumference of Joe's circle? (Use 3.14 for π.)

3. Darlene had made Joe's favorite lunch (potato pancakes) for him to enjoy while he was fishing. Joe *accidentally dropped one of the pancakes into the water.* If Bobbie and her remote control were on the boat, it could be one of her buttons: `take pancake and toss it into the water`. Does this have an inverse?

4. Joe likes ketchup on his pancakes. Darlene packed a large squeeze bottle containing 3 cups of Joe's favorite brand (Queen's Blood™). He used 7 fluid ounces on his first pancake. How much was left in the bottle? (1 cup = 8 fluid ounces.)

5. The set of all things that Joe is interested in might be described as $\{x \mid x \text{ is something that Joe is interested in}\}$. It also might be described as {eating, napping, goofing around}. Let's call that set *J*.
The set of all things that Darlene is interested in is {her nails, her hair, what's on TV, marrying Joe}. Let's call that set *D*. Find $J - D$.

Joe's favorite pancake topping

6. Darlene caught two fish while Joe was eating his pancakes. Both together weighed 8.3 kg. One weighed 5.8 kg. What did the other weigh?

7. $78 \times 0.78 = ?$

8. Round 3.007 to the nearest tenth.

9. Multiply 4.5 by a billion.

10. Which of these numbers—2, 3, or 5—divide evenly into 444111?

The Bridge

from Chapters 1–10 to Chapter 11

fifth try

---

1. Consider the function *take a number, add 13.5 to it, then divide it by π.* Does it have an inverse? If so, what is it?

2. You are taking a hike with some Cub Scouts. The hike is 3 miles long. After you have gone 67 feet, one of the Cub Scouts asks, "How much farther?" What is the answer to his question? (1 mile = 5,280 feet.)

3. One Cub Scout was 1.24 m tall. (A meter is a little more than a yard.) Another Cub Scout was 0.9 m tall. How much taller was the first than the second?

4. Round 645,000 to the nearest million.

5. What is the circumference of a circle whose diameter is 0.67 feet? (Use 3.14 for π.)

6. Which of these numbers—2, 3, or 5—evenly divides into 70000000000000004?

7. On the hike, you notice that Cub Scout Roger is eating jelly beans at the rate of 7.3 per minute. How many jelly beans will Roger eat during the 15-minute hike?

8. Let B = {Bobbie} and let P = {Fred, Alexander, Betty}. Find $B - P$.

9. Divide 0.7 by a million.

10. $6.7 \times 7.6 = ?$

Chapter Eleven
Dividing a Decimal by a Whole Number

At the beginning of the previous chapter we noted that Alexander had set the remote control on the floor and that Fred had turned off the light as he left the room. We had labeled those two facts as important.

So as Alexander, Betty, and Fred were striding, walking, jogging down the street toward the steak house, Bobbie was wandering around Fred's office in the dark.

You can guess what happened. She stepped on the remote.

One little foot hit one little button.

Fred and his two student friends finally arrived at the steak house. The neon sign in front was blinking on and off.

5100th Avenue Steak House

Betty said, "Divisible by five."

Alexander said, "Divisible by two."

Fred said, "Divisible by three, but not by nine."*

They entered the restaurant and sat down. Alexander rubbed his stomach. "It smells good in here. I hope they have large steaks." Betty was hoping they had moderately sized steaks. And Fred wasn't quite hungry yet.

The waiter came and handed each of them a menu. He was wearing a cowboy hat with a football team's logo on the side. He asked, "Waddle you have to drink?" (waddle = what will) Fred figured that was cowboy talk.

---

* Rule: If the sum of the digits is divisible by 9, so is the number.

Fred was about to say, "Sluice, please," when Betty put her hand over his mouth. She and Alexander had been going out to eat with Fred for years. And for years, each of them had ordered Sluice. It was one of the favorite drinks of KITTENS University students. Sluice, as everyone knows, is a clear lemon-lime soda not to be confused with Slice® / Sprite® / Storm® / Seven-Up® / Sierra Mist® / Squirt®, which all start with *S*. No straws are ever served with Sluice because it's so thick. If you put a spoon in a glass of Sluice, it floats. Sluice, the World's Sweetest Soft Drink℠, has a lot of sugar in it.

Betty has been concerned that Fred had not been growing. He was five years old and was only three feet tall. Without any parents, Fred had never been taught about eating good foods. This was one reason Betty and Alexander were taking Fred to a steak house—a little less sugar, a little more protein.

Betty said to the waiter, "We'll each have a milk, please."

"Okey-dokey," said the waiter. "I'll give you a moment to look at the menu while I get your drinks."

Everyone had a different reaction to the menu. Betty cringed at the misspellings. They had mixed up their homonyms: steak vs. stake. And, of course, *alot* should have been *a lot.*

Alexander figured he would get enough to eat.

Fred wondered if there was a child's menu he could look at.

The waiter came back and asked, "Waddle you have to eat?"

Alexander pointed to the only item on the menu.

Fred thought to himself 371.2 pounds of meat seems like a lot of meat, even for a family of 16. 371.2 divided into 16 equal parts would be. . . .

Fred took out a paper napkin to do the division.

When you divide a decimal by a whole number,* the procedure is easy. All you do is let the decimal float up into the answer.

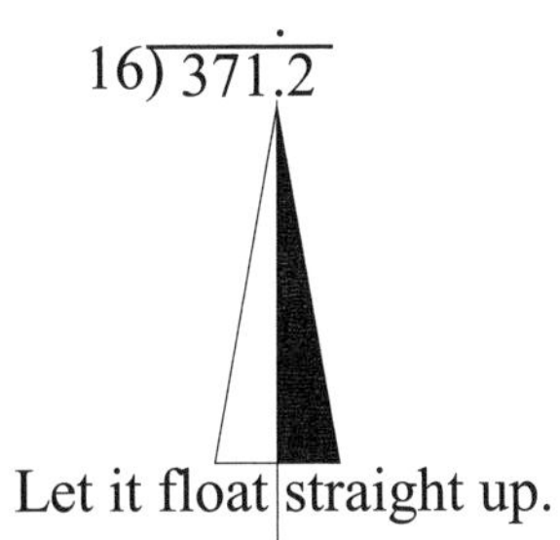

And then do the division the way you normally would.

```
      23.2
16) 371.2
   − 32
      51
    − 48
       32
     − 32
        0
```

To Fred's way of thinking, 23.2 pounds of meat for each member of a family of 16 was a lot of meat.

---

* The whole numbers = {0, 1, 2, 3, 4, 5, 6, 7, 8, 9, 10, 11, 12, . . .}

Your Turn to Play

1. What if you had a family of 32 to share the Half Cow Special. How much meat would each receive?

2. There is a Junior Half Cow Special that is not listed on the menu. It is designed for a family of seven. The Junior Half Cow Special is priced at $1,093.26. How much would that cost for each person?

3. $9\frac{1}{2} \times 5\frac{4}{5} = ?$

4. Are the natural numbers a subset of the whole numbers? (The **natural numbers** = {1, 2, 3, . . .}.)

5. $0.092 \times 0.007 = ?$

6. Round 777.077 to the nearest whole number.

7. Express 6° in minutes. (Recall, an angle of one degree is equal to 60 minutes.)

Complete solutions are on page 190

Chapter Twelve
When Division Doesn't Come Out Even

When Bobbie's little foot hit one little button on the remote control, it might have been something silly like `raise your arm`. But it wasn't. The button wasn't some housekeeping task like `dust the furniture`. She had stepped on the button that read `build a robot`.

- -

Fred realized that Alexander, Betty, and he weren't a family of sixteen people. There were only three of them, and they would be sharing 371.2 pounds of meat. A little drop of sweat formed on his forehead.

"I think we may have leftovers," Fred announced.

"Look!" exclaimed Betty. "They are putting on a show for us!"

A cowboy was herding a steer right through the restaurant.

"That's cool!" Fred said.

Alexander watched silently as the cowboy herded the steer into the kitchen.

The waiter came back to them with the three glasses of milk. "I forgot to ask. How would you like your steer cooked?"

Betty answered, "Medium, please. Could you tell me how long it will be till we're served?"

"It takes about five minutes," said the waiter.

Alexander was surprised. "Only five minutes. How do you do that?"

"We have a very large microwave oven."

Fred took a very little sip of his milk. He thought to himself *I wonder how I'm going to get hungry in the next five minutes. One little tiny crumb of hamburger would fill me up nicely.*

The lights in the dining room dimmed as the chef in the kitchen turned on the microwave. The microwave was as large as a drive-through car wash. In fact, that is what it used to be before they converted it into a microwave oven.

Fred thought *Maybe 253 students from the algebra class that I teach will show up at this restaurant. We could invite them to come to our table and share this giant meal. With 256 of us, we might be able to eat it all.*

He pulled out a paper napkin to compute how much of the 371.2 pounds of meat everyone would get if 256 people shared it.

$$\begin{array}{r}1.4\\256\,\overline{)\,371.2}\\-\underline{256}\\1152\\-\underline{1024}\\128\end{array}$$

Language lesson:

$$\text{divisor}\,\overline{)\,\text{dividend}}^{\;\text{quotient}}$$

quotient pronounced kwo-SHENT

Oops! It came out with a remainder of 128. It didn't come out even like the division problems in the previous chapter. What shall we do?

Back in Chapter 3, we had **The Trick of Adding Zeros.*** We will use that trick here. When it doesn't come out even, you add a bunch of zeros to the dividend. And keep dividing till the remainder is zero.

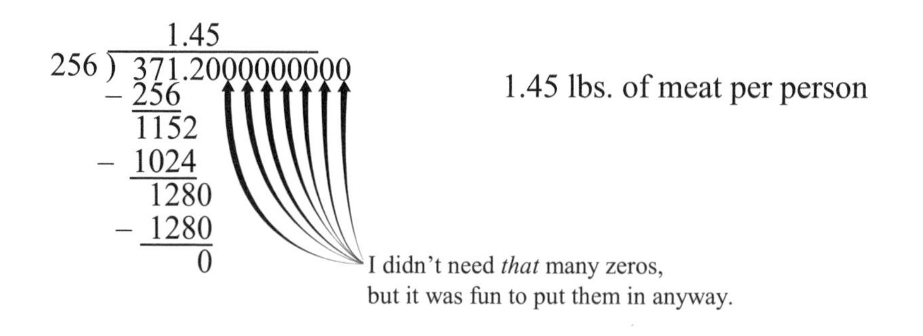

✶ "You can always add zeros on the *right* of the last digit that is on the *right* of the decimal."

Your Turn to Play

1. The 1.45 lbs. of meat when the half steer was shared by 256 people still seems like a huge amount. Suppose we divide the 371.2 lbs. of meat by 800 people. How much would each get?

2. We can now **change fractions into decimals**. Change $\frac{3}{8}$ into a decimal. (Divide 3.000000 by 8)

3. Change $\frac{1}{4}$ into a decimal.

4. **Changing decimals into fractions** is even easier. $0.127 = \frac{127}{1000}$

   Change 0.15 into a fraction.

5. Change 0.875 into a fraction.

6. $9\frac{1}{2} \div 5\frac{4}{5} = ?$

Complete solutions are on page 191

So Bobbie built a robot. That's what the remote control button said to do. She built a copy of herself and called her Berta. Berta Bot looked almost exactly like Bobbie. Together, Bobbie and Berta built Elizabot. Now they were triplets. They were each one-inch tall.

Five minutes passed. The waiter came out and told them, "Your order is ready—cooked medium, just as you asked."

Alexander put his napkin in his lap. Betty took a little sip of her milk. Fred didn't know what to do. In his life, he had never eaten anything except vending machine food, candy, and pizza. The only way that he had ever seen beef served was as hamburger. He was picturing a giant hamburger weighing 371.2 pounds.

None of the three pictured that the half steer would come on a single plate. A very large plate.

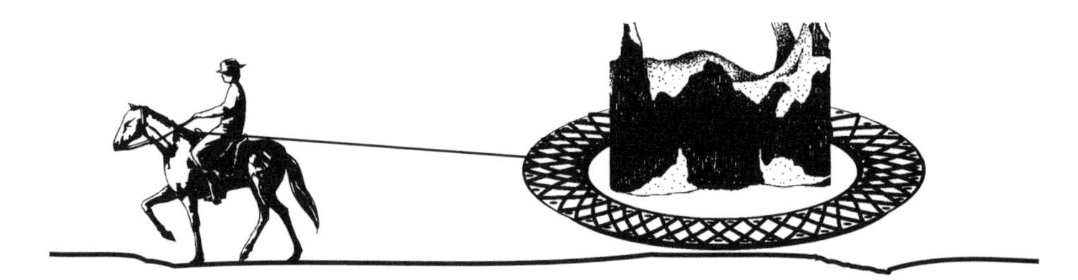

The waiter handed Alexander a large carving knife. The knife looked strangely familiar to Fred.*

---

* For those who have read *Life of Fred: Fractions*, this knife will be very familiar. This thirteen-pound beauty that was going to be a birthday present for Alexander had been taken by the doctor in the hospital. The doctor sold it on the Internet, and that's how the restaurant got it.

Alexander turned to Fred and asked, "What part would you like?"

Fred liked the sound of the word *part*. He was relieved that he wouldn't have to eat his fair share which would be one-third of the whole thing. Fred tried to think of the smallest part. "I'll have the tail, please."

Betty laughed. Fred knew a lot about math because he was a professor of math at KITTENS University, but he knew very little about eating beef. Betty showed Fred the carving guide on the back of the menu.

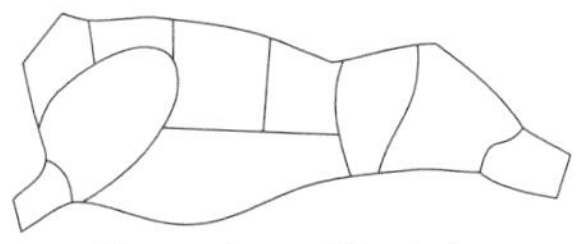

Carving Guide

She said, "You can choose." She read from the menu, "Top Blade Steak (also known as Book Steak or Butler Steak), Shoulder Steak (also known as English Steak), Chuck Arm Steak (also known as Round Bone Steak), Ribeye Steak (also known as Delmonico Steak), T-Bone Steak (also known as Porterhouse), Filet Mignon, Top Loin Steak (also known as New York Strip Steak or Ambassador Steak), Sirloin Steak, Tri-Tip Steak, Round Tip Steak, Round Steak (also known as Top Round London Broil), Skirt Steak (also known as Philadelphia Steak), Hanger Steak, and Flank Steak (also known as Jiffy Steak or London Broil)."

"What's the smallest?" asked Fred.

Alexander said, "Here, I'll give you a taste of filet mignon* and you can see how you like it." He cut off 0.4 of a pound and put it on Fred's plate. It smelled different than anything Fred had ever purchased from a vending machine. It smelled different than pizza.

Alexander calls that a taste. He gave me almost a half pound of meat Fred thought. Also new to Fred was the use of a knife and fork. He carefully watched Betty and tried to imitate her.

It wasn't working. Steak was tougher than pizza. Betty leaned over and cut his steak into six equal pieces.**

✶ (fill-LAY min-YON) It's from the French. It means "dainty slice."

✶✶ The RULES are that you may cut a child's meat up into lots of pieces, but on an adult plate you should only cut one or two pieces at a time and then eat them.

Fred wondered how much each of these pieces weighed. It was 0.4 divided into six equal parts. While the steak was cooling, he took out a paper napkin and did the division:

```
     0.0666
 6 ) 0.400000000
   – 36
       40
     – 36
         40
       – 36
          4
```

It wouldn't stop. The division was going to go on forever. Fred had five choices.

<u>Choice #1</u>: Fred could just keep on dividing. His answer would look like 0.06666666666666666666666666666666666666666666666666, and he would never have to eat any steak.

<u>Choice #2</u>: At some point, he could stop and use a remainder.

```
     0.0666 R 4
 6 ) 0.400000000
   – 36
       40
     – 36
         40
       – 36
          4
```

Writing a remainder is used less and less as you go on in mathematics.

<u>Choice #3</u>: At some point, he could stop and express the "leftovers" as a fraction.

```
     0.0666 4/6
 6 ) 0.40000          =   0.0666 2/3
   – 36
       40
     – 36
         40
       – 36
          4
```

$$\frac{4}{6} = \frac{2}{3}$$

<u>Choice #4</u>: Fred's quotient is a **repeating decimal**. One way to show that a decimal repeats is to put a bar over the repeating part.

0.0666666666666666 . . . can be written as $0.0\overline{6}$

Every fraction, when converted to a decimal, either terminates (stops) or becomes a repeating decimal. In the previous chapter, all the fractions were terminating decimals.

$\frac{1}{5} = 0.2$

$\frac{1}{4} = 0.25$

$\frac{1}{8} = 0.125$

$\frac{7}{8} = 0.875$

In this chapter, we encounter repeating decimals.

$\frac{1}{3} = 0.3333333333333\ldots$ which can be written $0.\overline{3}$

$\frac{1}{74} = 0.0135135135135135\ldots$ or $0.0\overline{135}$

<u>Choice #5</u>: Fred could round his answer. Suppose that in dividing 0.4 pounds of meat into six equal pieces, he wanted to know *to the nearest hundredth of a pound* how much each piece would weigh.

The procedure is to keep dividing until you go one more place beyond what you want, and then you round back.

Fred would have divided until he got 0.066 and then rounded it back to 0.07. In other words, he would have divided until he got to the thousandths place, and then he would have rounded it back to the hundredths place. In real life (and in engineering), rounding is often used.

Your Turn to Play

1. Fred's filet mignon was cut into 0.07-pound pieces. Convert 0.07 lbs. into ounces using a conversion factor. (1 lb. = 16 oz.)

2. Those $0.0\overline{6}$-pound pieces were too large for Fred's mouth. What if Betty had cut the 0.4-pound steak into 15 equal pieces? How much would each piece have weighed? Give your answer to the nearest thousandth of a pound.

3. Is $\frac{1}{7}$ a terminating or a repeating decimal? Or, more properly expressed, does $\frac{1}{7}$ terminate or repeat when it is written as a decimal?

4. Here's something to test your reasoning ability. We know from Chapter 5 that π is an unending, non-repeating decimal. We know that every fraction can be written as either a terminating or a repeating decimal. Therefore, what can we say about π?

5. Which is larger? $3\frac{1}{7}$ or π

($\pi \approx 3.14159265358979323846264338327950 28$)

$\approx$ means *approximately equal to.*

6. $2\frac{1}{3} + 4\frac{1}{7} = ?$

7. Will $\frac{20000000100060000}{3}$ be a repeating decimal?

Complete solutions are on page 192

Chapter Fourteen
Dividing by a Decimal

Bobbie, Berta, and Elizabot didn't know what to do next. When Bobbie and Berta had built Elizabot, they had used up the last of the tiny gears and motors in the watch parts bag. Berta suggested that they sit down and discuss what they should do next.

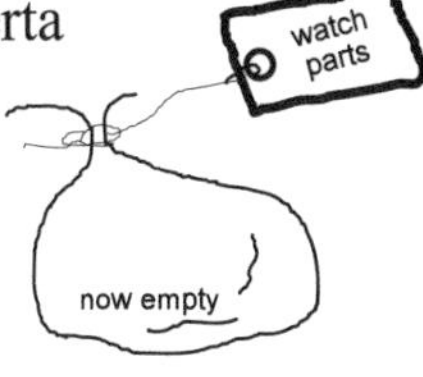

"Let's have a tea party," Elizabot said. She had gotten the idea from one of the books on Fred's bookshelves: Prof. Eldwood's *Modern Tea Parties*, 1853.

The three one-inch-tall robots all sat down near the electric outlet. Berta pretended to take out a teapot and some cups. Bobbie pretended to spread a tablecloth on a pretend table. Then all three recharged themselves.

Betty had cut Fred's filet mignon into six pieces, each weighing 1.12 ounces. Fred stared at one of the pieces. *This 1.12-ounce steak is a whole meal* Fred thought to himself. *I'm not sure I can finish this one piece.*

Alexander cut eight ounces of tri-tip for Betty and a pound of T-bone for himself. "Eat up. There's plenty for seconds and thirds," he announced. Fred had already put five of his six pieces of his steak into a doggie bag.

Fred took his knife and fork and cut his 1.12-ounce steak into ten pieces. Each piece now was 0.112 ounces. (Dividing by ten is easy. Just move the decimal one place to the left.) *Those are still too big to get into my mouth.* So he repeated the process by cutting each 0.112-ounce piece into ten smaller pieces. He had turned filet mignon into hamburger.

Using his spoon, Fred picked up one of those 0.0112-ounce pieces of meat. He couldn't use his fork, because those Fred-sized bites would fall right between the tines.

If Fred had cut those 0.0112-ounce pieces of meat any further, he would have invented *meat powder*. As it was, it would take about 1,429 Fred pieces to make up a pound.

Wait! Stop! I, your reader, don't follow that. How did you get 1,429?

I was about to show you.

So, show me.

Okay. All I did was divide 0.0112 ounces into a pound. I needed to find how many times 0.0112 went into 16 ounces. It's division.*

$$0.0112\,\overline{)\,16.000000}$$

You can't do that! All we know how to divide by are whole numbers like 9 or 3285. That's what you taught in Chapters 11, 12, and 13. We have never divided by a decimal.

I'll teach you.

Is it hard? I don't want "hard."

I think it's easier than learning to tie your shoelaces.

It took me two weeks to learn that.

This should take about twenty seconds. Here goes:

$$112.\,\overline{)\,160000.00}$$

What did you do?

I moved the decimal in the divisor over 4 places, and I moved the decimal in the dividend over 4 places.

0.0112 → 112.
16.000000 → 160000.00

Now you have a divisor that is a whole number.

---

* If you are ever uncertain—you don't know whether to add, subtract, multiply, or divide—here's a trick. Use really simple numbers first. If I ask you how many 2-oz. pieces of chocolate you can get from an 8-oz. bar, you would say—almost without thinking—"Four." Then go back and look at how you got your answer. You divided. Now when I want to find how many 0.0112-oz. pieces there are in 16 ounces, I divide.

So here are all the steps in finding out that it takes about 1,429 of Fred's 0.0112-oz. crumbs to equal a whole pound.

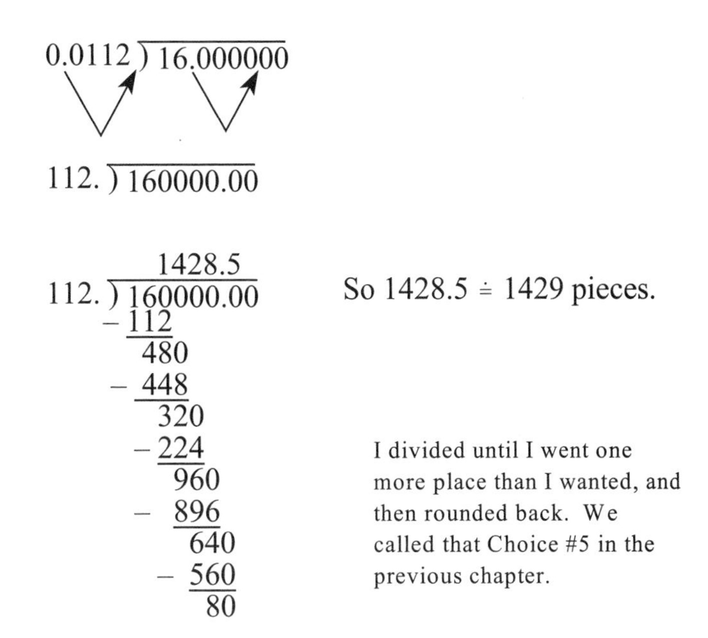

So 1428.5 ≐ 1429 pieces.

I divided until I went one more place than I wanted, and then rounded back. We called that Choice #5 in the previous chapter.

Your Turn to Play

1. ***Wait! Stop! I, your reader, won't let you have a*** *Your Turn to Play* ***yet. Before you start asking your questions, I have a question for you, Mr. Author.***

But we're right in the middle of a *Your Turn to Play*! How can you stop me?

Easy. I'm going to end this and skip down below.

.......COMPLETE SOLUTIONS.......

1.

See! We're no longer in the *Your Turn to Play* ***stuff. Now answer my question: Why does this trick of yours work? Why is it true? What gives you the right to move the decimal in the divisor and the decimal in the dividend?***

This is not fair! I'm the author of this book. Don't I get to write what I want?

No.

Why not?

Because I'm holding this book in my hands right now. You aren't. Readers always get the final say. If I want to draw on these pages, I can. Here's my drawing of a mouse.

See, I can draw just as good as you can.

If you want to get on with this chapter, you had better answer my question: What gives you the right to move those decimals over?

I can draw a pretty good mouse.

Please. Just answer my question.

Okay. We started with

$$0.0112\overline{)16.000000}$$

That means the same as

$$16.000000 \div 0.0112$$

We can express division as a fraction

$$\frac{16.000000}{0.0112}$$

It doesn't change the value of a fraction if you multiply numerator and denominator by the same number. We know that $\frac{3}{4} = \frac{3 \times 2}{4 \times 2} = \frac{6}{8}$

Multiplying top and bottom of $\frac{16.000000}{0.0112}$ by 10,000 yields $\frac{160000.00}{112.}$

which is the same as $112.\overline{)160000.00}$

Thank you. You can do the *Your Turn to Play* ***now if you wish.***

Your Turn to Play

1. $0.25\overline{)80.}$ This is the same as asking how many quarters there are in \$80.

2. Because $0.25 = \frac{25}{100} = \frac{1}{4}$ please do the previous problem using fractions.

$80 \div \frac{1}{4} = ?$

3. How many millionths are there in a million?

4. $4\frac{1}{7} - 2\frac{1}{3} = ?$

5. Change $\frac{5}{8}$ into a decimal.

6. To go from the diameter of a circle to its circumference, you multiply by pi. To go from the circumference to the diameter, you divide by pi. Suppose the circumference of a circle is 23 miles. To the nearest mile, what would be its diameter? (Use 3.14 for π.)

7. $\pi^2 = ?$ π^2 means $\pi \times \pi$. π^2 is read "pi squared." (Use 3.14 for π.)

Complete solutions are on page 192

Chapter Fifteen
Bar Graphs

Berta poured some imaginary tea into Elizabot's imaginary teacup and asked if she would like some lemon in her tea. Elizabot nodded and she squeezed a couple imaginary drops into her cup.

"I have this overwhelming urge to build a robot," Bobbie announced.*

"Me too," said Elizabot.

Berta was the sensible one. She informed the other two, "But we're out of watch parts. We used the last little gear and motor in making you, Elizabot. All we have left is the big junk."

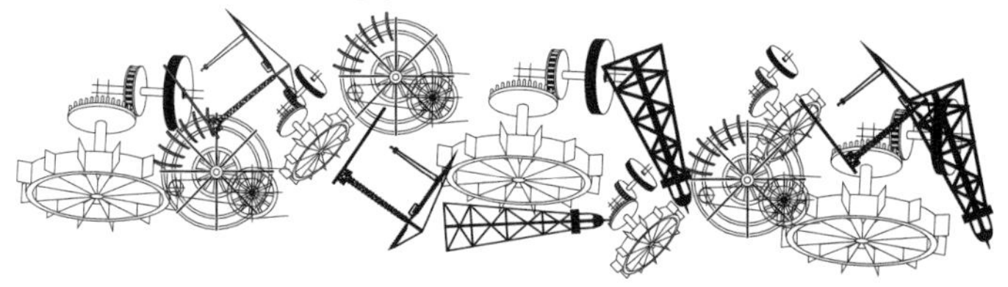

There was silence for a moment. Then all six electric eyes twinkled.

Fred had had a little bite of marshmallow for lunch. For dinner, so far, he had had a little sip of milk. In his spoon he held 0.0112 ounces of filet mignon. That bit of beef never made it to Fred's lips.

Food and *Fred* may start with the same letter. They may end with the same letter. But somehow, they never seem to get together.

The ground *shook*. It *shook* again. It *shook* again.

"Earthquake?" asked Betty.

Fred put down his spoon.

"No, it's too regular," said Alexander. "It sounds like a giant monster walking the earth."

A pretty good guess.

---

* The `build a robot` button on the remote control that Bobbie had stepped on was stuck in the down position.

They ran outside to look.

Bobbie, Berta, and Elizabot had built a big robot. They had started by using all the big parts that were in Fred's office. While they were running around assembling the pieces, Bobbie, again, stepped on the remote control. She mashed down the button marked `bigger`.

When they had built a robot that was about six feet high, they were dissatisfied. Something inside each of them said, "Bigger." They tore it apart and took the pieces outside. When they ran out of parts from Fred's office, they used whatever they could find: power poles, car fenders, mailboxes, and satellite dishes.

When they were done, Bobbie took the remote control in her hand. She noticed that the `build a robot` and the `bigger` buttons were stuck in the down position, and she popped them up. Suddenly, the urges were gone.

- -

Alexander stared at the monster robot. His mouth was open.

Betty took some pictures.

Fred drew a bar graph. He thought it would be great for teaching math in his classroom.

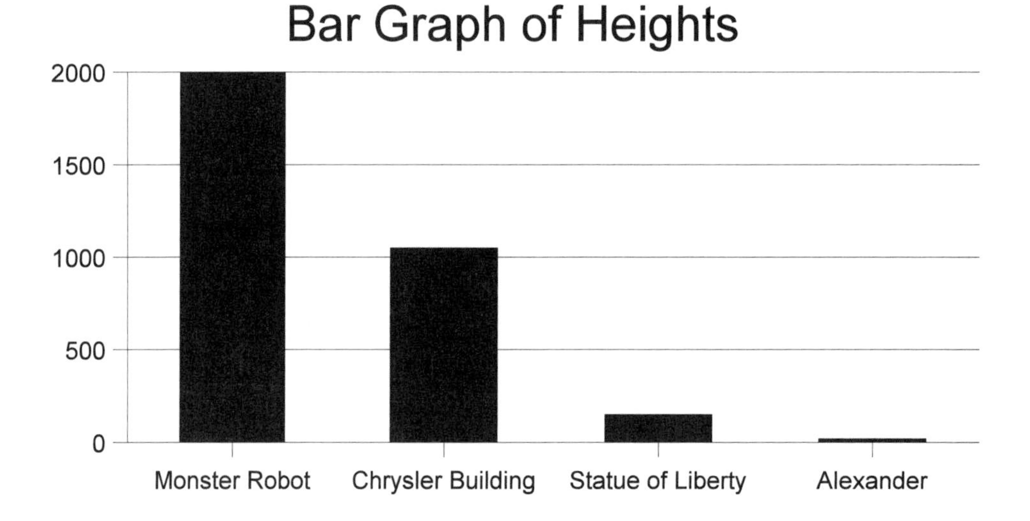

Looking at the bar graph, you could tell that the robot is 2000' tall. The Chrysler Building is about 1000' tall. The Statue of Liberty is about 200' tall. It's hard to guess how tall Alexander is by looking at the graph.*

Fred had drawn a vertical bar graph.

✶ The actual numbers are: Chrysler Building, 1050 feet; Statue of Liberty, 151 feet; and Alexander, 6 feet.

He could have drawn a horizontal bar graph.

Horizontal Bar Graph of Heights

Alexander
Statue of Liberty
Chrysler Building
Monster Robot

0 500 1000 1500 2000

But he didn't. The vertical bar graph gave a feeling of height that a horizontal bar graph didn't convey. However, if he were doing a bar graph of how far different vehicles could go on a gallon of gas, then a horizontal bar graph might have looked better.

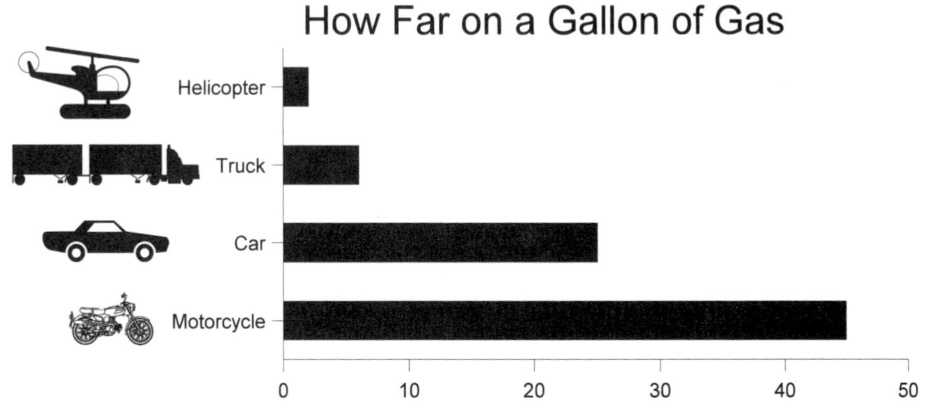

You'll notice that on the bar graph at the top of this page, the horizontal axis went from 0 to 2000. On the gallon-of-gas bar graph, the horizontal axis went from 0 to 50. When you draw a bar graph, you first look at the data and they* will tell you what numbers to stick on the numerical axis.

---

* *Data* is plural. You say things like, "The data are. . . ." One item of data is called a datum.

Your Turn to Play

1. Alexander had eaten 2400 calories today. Betty, 1350. Fred, 86. If you were drawing a bar graph to show the number of calories that they had eaten, how might you label the numerical axis? From 0 to what?

2. How many Statues of Liberty (151 feet) would you have to stack up on top of each other to equal the height of the Chrysler Building (1050 feet)? Give your answer to the nearest whole number.

3. $7.8^2 = ?$ 7.8^2 means 7.8×7.8 and is read "7.8 squared."

4. Change $\frac{1}{8}$ into a decimal.

5. The whole numbers are W = {0, 1, 2, 3, . . .}. The natural numbers are N = {1, 2, 3, 4, . . .}. The natural numbers are sometimes called the counting numbers. Find W – N and N – W.

Complete solutions are on page 193

The Bridge

from Chapters 1 – 15 to Chapter 16

first try

Goal: Get 9 or more right and you cross the bridge.

1. The restaurant claims that their Junior Half Cow Special is designed for a family of seven. It weighs 124.6 lbs. How much would that be per person (assuming they all ate the same amount)?

2. Express 0.8° in minutes. (One degree of angle equals 60 minutes.)

3. Change 0.73 into a fraction.

4. Change 44.73 into a mixed number. (Numbers like $9873\frac{1}{8}$ or $2\frac{1}{5}$ are called mixed numbers.)

5. If Betty had cut Fred's 0.4 pound steak into 7 equal pieces, how much would each piece have weighed? Give your answer to the nearest thousandth of a pound.

6. Express $\frac{1}{6}$ as a repeating decimal. (Use the notation that puts a bar over the repeating part.)

7. Express $\frac{2}{7}$ as a repeating decimal. (Use the notation that puts a bar over the repeating part.)

8. If it takes a waiter 0.097 hours to set a table, how many tables could he set in twenty hours? (Hint: Use a conversion factor like the one on page 191. $\frac{1 \text{ table}}{0.097 \text{ hours}}$) Give your answer to the nearest table.

9. Name a whole number that is not a natural number.

10. A decade is ten years. How long is a decade minus one week? (Assume one year = 52 weeks.)

The Bridge

from Chapters 1 – 15 to Chapter 16

second try

1. When Joe and Darlene went fishing, Darlene had packed Joe's favorite lunch (potato pancakes). Joe took the first pancake, which weighed 450.6 grams, and ate it in three equal bites. How much did each bite weigh?

2. Darlene made him some ambrosia for dessert. (The cookbooks will tell you that it is made from oranges, shredded coconut, and sometimes pineapple.) Joe didn't like it and secretly dropped handfuls of it over the edge of the boat when Darlene wasn't looking. It took eight handfuls in order to get all of it dumped overboard. (One handful = 7.8 ounces.) How much did Joe get rid of?

3. In order to dry off his finger, Joe used his finger on the side of the boat to write the largest number he could think of. He wrote 222222222222222222. He wrote eighteen *2*s. Is that number evenly divisible by 9?

4. Change 7.8 oz. into a mixed number. (Numbers like $9873\frac{1}{8}$ or $2\frac{1}{5}$ are called mixed numbers.) Remember the rule that you always reduce a fraction as much as possible.

5. Because it took eight handfuls for Joe to get rid of all of the ambrosia, he got rid of $\frac{1}{8}$ of it with each handful. Express $\frac{1}{8}$ as a decimal.

6. Darlene spent 144 minutes slicing the oranges for the ambrosia. She sliced at the rate of 0.06 of a minute per slice. How many slices did she make?

7. Round your answer to the previous problem to the nearest thousand.

8. Joe caught 9.04 pounds of fish. They lost 1.5 pounds because they dried out in the sun. What was their weight after drying out?

9. The largest fish that Joe caught was 6.7 inches long. After Darlene chopped off the head (1.3") and the tail (1.5"), how long was it?

10. Joe took the head and the tail and played with them as if they were puppets. How much taller was the tail than the head?

The Bridge
from Chapters 1 – 15 to Chapter 16

third try

---

1. Darlene decided to sew a very pretty dress that she could wear if Joe ever asked her to go out dancing. The first step was to get some thread. Rather than just go out and buy a spool of thread, she wanted to spin the thread on a spinning wheel. Then Joe would be really impressed.

She had heard that a sheep would yield enough wool to make 69.3 yards of thread.

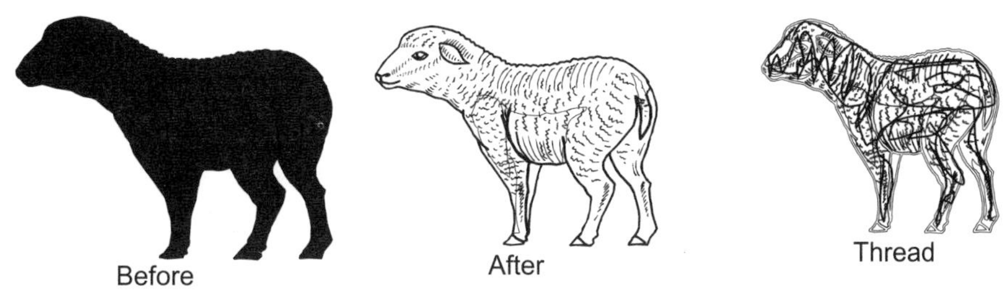

How many sheep would Darlene need in order to get 3257.1 yards of thread?

2. Change 3257.1 yards into a mixed number. (Numbers like $9873\frac{1}{8}$ or $2\frac{1}{5}$ are called mixed numbers.)

3. Darlene and four of her girl friends carried the first sheep into her bedroom so that she could shear the wool off of it. The sheep weighed 184 pounds. Assuming that they all carried an equal amount, how much did each of the five carry?

4. Express $\frac{1}{5}$ as a decimal.

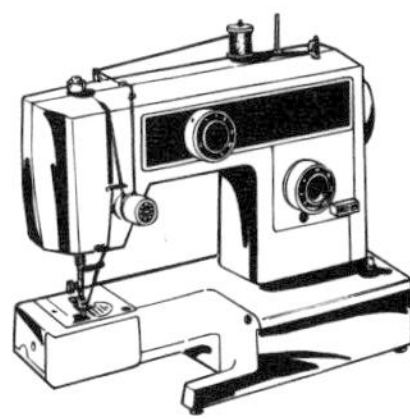

5. If Darlene can use up 70.3 yards of thread in an hour, how long will it take her to use up the 3257.1 yards? (Give your answer to the nearest tenth of an hour.)

6. Round 3257.1 to the nearest hundred.

7. How much did the rental sewing machine cost (50.8 hours at \$2.35/hr)?

8. $56.21 \div 0.07 = ?$

9. $56.21 - 0.07 = ?$

10. $56.21 \times 0.07 = ?$

The Bridge

from Chapters 1 – 15 to Chapter 16

fourth try

---

1. Change 6.78 into a mixed number. Remember to reduce fractions as much as possible.
2. Is 923492369293469292346990747293472392 0 evenly divisible by 5?
3. 251.43 ÷ 5.1 = ?
4. Five sailors who were visiting New York decided to go and see the Statue of Liberty. They got into a cab for the 20.5-mile trip. The cab charged $4.02 per mile. What was the total bill?
5. Continuing the previous problem, to the nearest cent, what was the bill for each of the five sailors?
6. Continuing the previous problem, each sailor paid $20. Part of that went to pay for the bill, part of that was the tip. What was the total tip received by the cab driver?
7. One of the sailors chipped off one-twentieth of an ounce from the statue. He wanted a souvenir. (One of the Ten Commandments talks about stealing and indicates that you are not supposed to do that.) Express $\frac{1}{20}$ as a decimal.
8. Assume the statue weighs a billion ounces. (A billion is 1 followed by nine zeros.) Suppose everyone did what that sailor did. How many people would it take to chip away the whole Statue of Liberty?
9. The sailors had to get back to their ship by midnight. It was now 5 minutes and 20 seconds past 5 PM. How long did they have left?
10. In order to travel the 5.08 miles back to the ship, they figured that it would take 23.7 minutes. But there was a traffic jam, and it took three times as long to get back to the ship. How long did it take them? Give your answer in hours and minutes (rather than just in minutes).

The Bridge

from Chapters 1 – 15 to Chapter 16

fifth try

---

1. If the Junior Half Cow Special cost $1,093.26, how much would a dozen (12) of those Junior Half Cow Specials cost?

2. Continuing the previous problem, if 21 people split the cost of a single Junior Half Cow Special, how much would each pay?

3. Betty eats 1.7 times more slowly than Alexander. If Alexander takes 8.9 minutes to finish his steak, how long will it take Betty?

4. Change 8.9 minutes into a mixed number.

5. Change 8.9 minutes into seconds.

6. If the diameter of a circle is 0.009 miles, what is its radius?

7. Here is a nine-digit number with one of the digits missing:

879?205345.

What must that missing digit be in order for the number to be evenly divisible by 9?

8. $6.4\overset{?}{\overline{)5.7984}}$

9. Subtract one inch from 50 yards.

10. Express $\frac{4}{7}$ as a repeating decimal. (Use the notation that puts a bar over the repeating part.)

Chapter Sixteen
Prime Numbers

Fred felt something on his foot. He looked down and saw Bobbie standing on his right shoe. It's a well-known fact that one-inch tall robots like to be around people like Fred that are only three feet tall. Then they don't feel so short.*

"Well, what do you think?" Bobbie asked.

"About what?"

" 'Bout the big bot we built," Bobbie answered, using as much alliteration as she could muster.

"You built that!" exclaimed Fred.

"Well, you told us to. Look." Bobbie pointed to the remote control that was leaning up against Fred's leg. "There's the remote control, and the two buttons you had pushed down: `build a robot` and `bigger`."

Fred picked up the huge remote control—the one with 168 buttons. "Will this control the giant robot you built?"

"No. That just controls me and my two little sisters, Berta and Elizabot."

Fred looked down and noticed that Berta and Elizabot were standing on his left foot, playing with his shoelace. "But . . . but . . . but how do you control her," Fred sputtered as he pointed to the giant robot.

"You are so silly, Freddie," said Bobbie. "That's not a girl robot. It's a boy robot. Can't you tell? His name is Roger Robot. Girl robots drink electricity, and boy robots like gasoline."

Roger slowly looked around and spotted a gas station. He pressed the premium button and put the nozzle between his lips. The flow was too slow for Roger's taste. He sucked, and the station collapsed.

"Table manners!" Bobbie chided him.

Fred sensed trouble was coming. Perhaps, *sensed* is not the right word. The government was moving in. Soldiers were pouring out of

---

* You want proof? When is the last time you saw a whole bunch of one-inch tall robots congregating around someone who is six feet tall?

trucks. Tanks were rolling down the street. Fighter jets were circling overhead.

He picked up a copy of the newspaper.

THE KITTEN Caboodle

The Official Campus Newspaper of KITTENS University Monday 5:14 p.m. Edition 10¢

exclusive

Death Monster Attacks!

KANSAS: A giant war machine has invaded our campus. According to a high-placed government source, "This is obviously a terrorist attack by some Communist nation. We can expect hundreds of these infernal weapons to show up at any moment." The government official has asked to remain anonymous.

Reports have come in that this death monster has completely destroyed a local refinery. It is obviously attempting to cripple our access to energy.

Mabel Muffins, head of KITTENS home economics department, was an eyewitness. She reports, "The monster must be at least 5000 feet tall. I saw at least one of them, and there might have been another couple of them. It was hard to tell."

file photo

In an exclusive interview, Capt. McTuff, Kansas Coast Guard, has stated, "Without a doubt, we are not sure what we are facing. We are ready to go nuclear."

Fred noted a couple of errors in the paper. He thought I guess the paper wouldn't sell as well if they knew that it was just a toy made in a math professor's office.

Then the cavalry came. Fred counted 17 men on horseback.

"Okay men. This monster is a threat to our national security," the cavalry leader shouted. "Divide into two equal groups and surround the monster."

"Boss. We can't do that," shouted one cavalryman. "There's seventeen of us."

"Okay. Then divide into three equal groups and surround the monster," the leader shouted.

"No can do.✶ There are seventeen of us."

Fred was starting to giggle as he listened. He knew that 17 was a **prime number**. The leader wasn't going to be able to split them into any number of equal groups. No number divides evenly into 17 except 1 and 17.

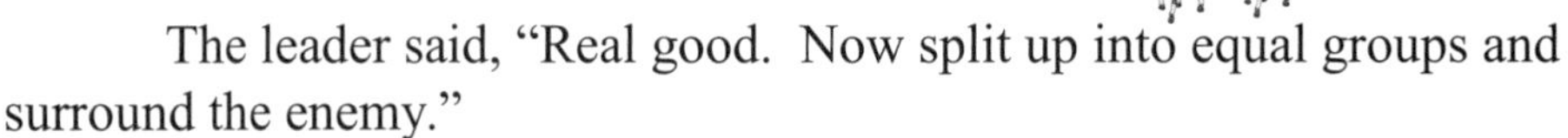

Two more men on horseback arrived.

The leader said, "Real good. Now split up into equal groups and surround the enemy."

"Boss. There are now 19 of us. That's still a prime number."

The leader thought for a moment, and then he ordered one of the cavalrymen to go and buy some doughnuts. Now with 18 left, they could split into equal groups.

✶ That's military talk. It's much shorter (and tougher sounding) than saying, "We can't do that."

Your Turn to Play

1. Numbers, like 18, that are not prime are called **composite numbers**. Eighteen cavalrymen could be split into two groups of 9. Name three other ways that they could be split.

2. Some of the natural numbers, {1, 2, 3, 4, 5, . . . }, are prime. Thirty-seven is a prime number, because it has exactly two divisors—1 and 37. Having exactly two divisors is the definition of a prime natural number. Those natural numbers that have three or more divisors are composite numbers. Not every natural number is either prime or composite. There is an exception. Name the natural number that is neither prime nor composite.

3. There is only one prime number that is even. Name it.

4. List all the primes that are less than 60.

5. A common mistake in doing the previous problem is to say that 51 is a prime number. That is because most people don't know their 17-times tables. $51 = 3 \times 17$. But even without knowing the 17-times tables, there is an easy way to see that 51 is composite. How?

6. $4\frac{1}{7} \times 2\frac{1}{3} = ?$

7. $1090 - 1.09 = ?$

8. **Consecutive numbers** are numbers that are right next to each other, such as 77, 78, and 79. Find five consecutive numbers that add up to 135.

9. Is it possible to name two consecutive numbers greater than 100 that are both prime? If so, then find an example. If not, explain why you can't find two consecutive numbers greater than 100 that are both prime.

Complete solutions are on page 193

One of the cavalrymen rode up to Roger Robot and hit him with his sword. His sword broke. *Why are they attacking him?* Fred wondered. *Roger doesn't look like a mean robot. He's just big.* Just then, a tank fired a shell at Roger's leg.

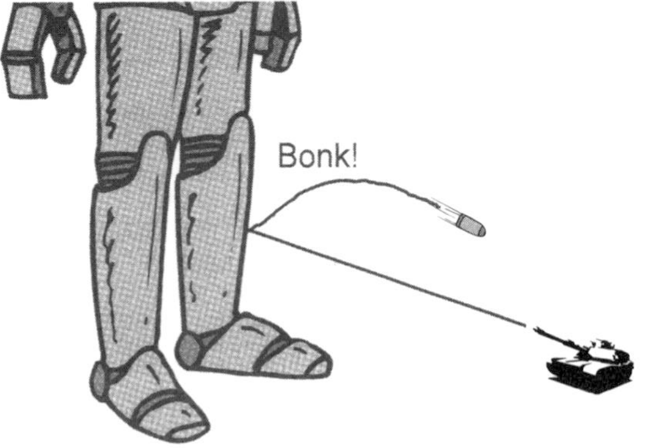

Fred was getting a little concerned. He asked Bobbie, "Well, if the huge remote control doesn't command Roger, what does?" Fred was thinking *I hope she doesn't say that nothing controls him.*

Bobbie held out her three-tenths-of-an-inch arm and showed Fred a tiny remote control. "This does it. Don't you think it's cute? I can make Roger do anything with this little remote."

"Give it to me, please," Fred said.

"No. Roger is my robot. I won't give up the control." Bobbie stamped her little foot.

Fred took the huge remote control—the one that controls the one-inch-tall robots—and examined the 168 buttons. He found the one he wanted: `hand over the tiny remote control`. He pushed that button.

Bobbie handed Fred the tiny remote control. She wasn't very happy. Fred looked over the buttons on the tiny remote control:

- `go shopping`
- `conquer the world`
- `sing "My Country 'Tis of Thee"`

- sit down and play
- write a history of polyester
- register to vote
- give Fred Gauss a big kiss
- explore Hudson Bay
- build an aviary
- work on the Goldbach conjecture

Fred was amazed. Bobbie really had thought of almost everything to put on the tiny remote control she had built for Roger. Some of the buttons Fred hoped would never be pressed.

But he did want the government to stop attacking Roger. Fred pressed the sit down and play button. *Maybe that will show them that Roger isn't like King Kong* Fred thought.

Fred also pressed the be careful where you sit! button. Roger looked around, found the KITTENS football field, and sat down. When the tanks rolled onto the field, Roger picked them up and played with them as any little boy might play with toy trucks.

Your Turn to Play

1. Many readers thought that Fred was going to press the work on the Goldbach conjecture button because that is the title of this chapter. Being very interested in mathematics, Fred was really tempted to press that button. He would have been delighted to see what Roger could find out about the Goldbach conjecture, but Fred was more concerned that Roger not be killed.

Wait! Stop! I, your reader, have a dozen questions jumping around in my head. First, what does "conjecture" mean? Second, who is Goldbach? Third, what is the Goldbach conjecture? Fourth, how could Roger register to vote?

That's a lot of questions to answer. First, a conjecture is something you think is true but aren't absolutely certain that it is true. So when Christian Goldbach wrote a letter to Leonhard Euler back in 1742, he mentioned that he had happened to notice that every even number greater than 2 could be written as the sum of two primes. $4 = 2 + 2$ $6 = 3 + 3$ $8 = 5 + 3$ $10 = 5 + 5$ Now, please continue this up to $50 = 37 + 13$.

2. ***Wait! You didn't finish answering my questions.***

You were asking how Roger could register to vote. I'm not sure what would happen if you pressed that button. Because Roger is about 30 minutes old, they probably wouldn't let him register.

I have another question. Is this little thing in a note to Euler called the Goldbach conjecture?

Yes. Goldbach didn't know whether it was true or not.

Well . . . is it true? Can every even number greater than 2 be written as the sum of two primes?

Goldbach didn't know. Euler [pronounced oiler] didn't know. No one in the 1700s ever figured out whether it was true.

And no one in the 1800s. And no one in the 1900s. And no one in the 2000s.

Well, is it true?

A lot of mathematicians have worked on this over the centuries. This simple little question—can every even number greater than 2 be written as the sum of two primes?—and—how can I break the news to you?

Spit it out! Is it true or not?

We don't know. Heaven only knows. It's called an **open question** in mathematics. Great fame awaits you, my reader, if you can either prove that it is true, or you can find some big even number like 2394239452952926394234933492369669544340530 6 which can't be expressed as the sum of two primes.

To continue the story a bit, Goldbach also happened to notice that every number greater than 5 could be written as the sum of three primes—not just the *even* numbers, but *every* number. 8 = 2 + 3 + 3
16 = 11 + 3 + 2 19 = 11 + 5 + 3 Show this is true for 20, 25, and 30.

Incomplete solutions are on page 194

Chapter Eighteen
Area of a Circle

Have you ever looked in the sports section of your newspaper and found that the KITTENS University football team was not even mentioned? How frustrating that must be. They have pictures of the winners and the losers. But where is the KITTENS team? In response to the hundreds of letters that Polka Dot Publishing has received asking about football at KITTENS, we dedicate this chapter to the team.

Oops!

Football had never been very important at KITTENS University. Many other universities glory in their football teams. In contrast, at the **K**ansas **I**nstitute for **T**eaching **T**echnology, **E**ngineering, and **N**atural **S**ciences, it is education (Heavens!) that is most important.

In the first four years that Fred taught at KITTENS, they did have a football team. The seven students who signed up for the team practiced every week during the fall semester under the guidance of Mabel Muffins.

Years ago, Betty and Alexander took Fred to one of the games so that he could say that he had been to a game once in his life. Fred tried to get into the spirit of it all, and when the opposing team had declared a timeout, Fred said, "Rah." Alexander told Fred that you're supposed to yell "Rah!" and not just say it. Fred liked timeouts. It meant that there would be two minutes without anyone knocking anyone else down.

But what happened today on the KITTENS football field ended the university's involvement in the sport forever. Today was the day that the school lost its football field.

Roger Robot was sitting in the middle of the field playing with his "toy tanks." The cavalry was throwing rocks at him, but he didn't notice.

The president of the United States was in his oval office doing secret things we aren't supposed to know about.

The vice president was off hunting.

The members of Congress were off on congressional investigations in Las Vegas and Hawaii.

That left Capt. McTuff in charge of our nuclear arsenal, and it is not hard to guess how he proceeded.

It was a small atomic bomb. They dropped it into Roger's lap. He picked it up. It exploded.

The only thing that was left was the tiny remote control in Fred's hand. Roger was gone. The field was gone. There was just a big circular hole where Muffin's Seven used to play.

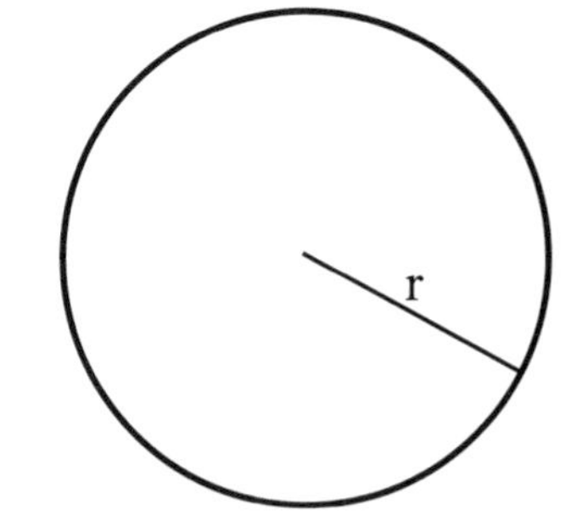

The radius of that circle was r = 45 feet.

The area of that circle is $A = \pi r^2$.
This could be written as $A = \pi \times r \times r$.

Your Turn to Play

1. What is the area of a circle that has a radius of 45 feet? (Use 3.14 for π.)

2. Round your answer to the previous problem to the nearest square foot.

3. Suppose you have a square that is 10 feet on each side. What is its area? (The area of a square is $A = s^2$ where s is the length of a side.)

10'

4. Continuing the previous problem, suppose you draw a circle inside the square.

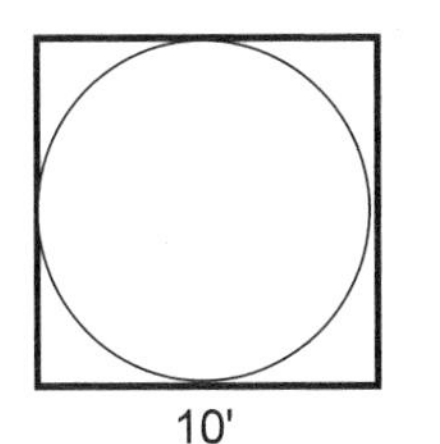

What would be its area?

5. The area of a rectangle is its length times its width.

   $A = \ell w$

   If $\ell = 78$ miles and $w = 37$ miles, what is the area of the rectangle?

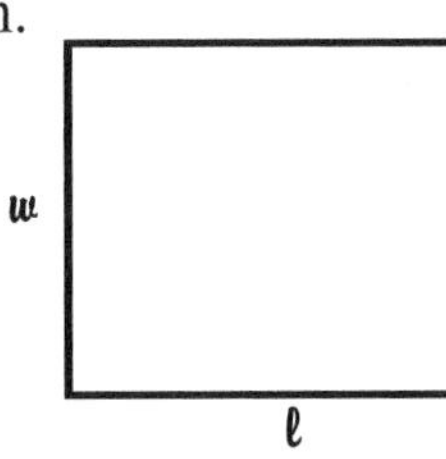

6. Continuing the previous problem, suppose you draw the largest possible circle inside that rectangle. What would be the area of that circle? (Use 3.14 for π.)

7. Continuing the previous problem, round your answer to the nearest tenth of a square mile.

8. $4\frac{1}{7} \div 2\frac{1}{3} = ?$

9. In problem 5 we wrote $A = \ell w$. The length and width were written in *cursive letters* rather than in this font (this is Times New Roman). Why do you think we did that?

Complete solutions are on page 194

Chapter Nineteen
Dollars vs. Cents

Things were pretty quiet after the explosion. A fine metallic dust settled over everything like a silent rain. It was half Roger and half football field dirt. People with small noses did a lot of sneezing. Fred was blessed with an ultra-filtration-system nose. He hardly noticed anything at all.

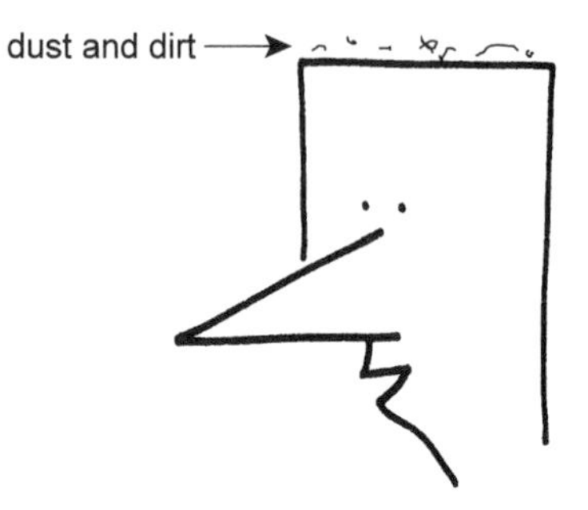

Except his scalp itched a little.

The cavalry headed off to the barn. Capt. McTuff wrote up his report. It was in the form of a memo to the president. McTuff stated in the first paragraph that he had saved the United States from almost certain destruction. In the fourth paragraph he mentioned that it was necessary to "do a little nuking of the enemy."

Mabel and her seven would start baseball practice next week.

Alexander and Betty were still hungry. Betty urged Fred to join them back inside the steak house, but Fred said that he had had enough to eat and was going to walk home. Bobbie, Berta, and Elizabot followed Alexander and Betty into the steak house. The three girl-robots wanted a little electricity for dessert.

It was time for a walk in the twilight hours.* Fred had a lot to think about. The idea of owning a bicycle no longer had the same attraction that it did at the start of the day. As he walked down the street, he briefly thought of getting a helicopter instead of a bike. Then he could get to class really fast. He could have the university build helicopter landing pads near his office and the places where he would like to land. After a few moments that idea drifted off into the evening air, joining his memories of Billy Bug and of a trip he had once made through a car wash while sitting

---

✶ What everyone calls a crepuscular perambulation (per-am-byuh-LAY-shun). To perambulate is to take a stroll, often in a public place. If I hadn't told you what *perambulation* meant, you might have had to look it up in a dictionary.

on the hood of the car. When you get old, you get a lot of "leftover" memories like that.

He was walking by the C.C. Coalback Toy Store. He stopped and looked at the sign in the window.

Fred thought *I don't have many toys.*✶ *I think I can afford a toy helicopter. It's probably cheaper than owning a real one.* He walked inside.

The woman at the cash register greeted him, "Hi kid. Where are your parents? Don't touch the toys unless you're going to buy them. Are your hands clean? Do you have any money on you? Don't take all day looking around."

Fred had never been to a toy store before. *I guess they don't want you to fool around* he thought. *Okay, I'll get right to the point.*

He walked up to the cashier and said, "Please. I'd like to have two of the helicopters you have advertised on your window sign."

✶ How many toys does Fred own? That's the same as asking what is the smallest natural number. (N = {1, 2, 3, . . . }) Fred has a sleeping bag and a lot of books. The microscope and the electronic scale mentioned earlier in this book belonged to the science department at KITTENS. His only toy is a doll named Kingie that he received from the cashier at King of French Fries when Fred was four days old. You'll learn more about Kingie in Chapter 6 of calculus.

Kingie

"Sure. Everyone would like to have a couple of helicopters," she snarled. "You got the money kid?"

"Of course," Fred replied. He placed a penny on the counter.

Your Turn to Play

1. "You trying to be funny, kid?" the cashier said. "Two helicopters will cost you a dollar." If that was true, her sign was incorrect. What should have the sign read?

2. Convert .50¢ into a fraction.

3. If the sign were correct, and you could buy a toy helicopter for .50¢, how many could you buy for a dollar? (Divide 100¢ by .50¢.)

4. Write each of these using the "¢" sign. $0.42 $7.04 $800

5. Write each of these using the "$" sign. 33¢ 598¢ 6¢

6. A 16-inch pineapple pizza costs $10. What is the area of that pizza? (Pizzas are measured by their diameters. Approximate π by 3.14.)

7. Continuing the previous problem, how much does it cost per square inch? Round your answer to the nearest cent. (If you are not sure whether to add, subtract, multiply, or divide, create a very simple problem with easy numbers first. For example, suppose that 8 square inches cost 16¢. Then one square inch would cost 2¢. Now, how did you get the 2¢? You divided the square inches into the price.)

Complete solutions are on page 195

Chapter Twenty
Pie Charts

Fred took his penny off the counter. As he walked down the aisle of the toy store, the cashier warned him, "Stay in the center of the aisle. Don't get too close to the toys. There's no gum chewing in the store. Keep your hands in your pockets."

Fred decided that he liked libraries more than toy stores. At least, libraries are better than the C.C. Coalback Toy Store. In libraries, they will let you touch the books. You can take them off the shelves and look at them. You can even borrow them and take them home . . . for free!

He left the toy store. Because this was the first toy store that he had ever been in, Fred didn't know that most toy stores don't treat their customers as badly as the C.C. Coalback Toy Store does.

Next week, it would be out of business.

Did you know that forty percent (40%) of all toy stores that treat their customers this badly go out of business within a month? No one likes to shop there.

Wait! You did it again, Mr. Author.

I know what you're going to say. You're going to tell me that the title of this book is *Life of Fred: Decimals and Percents*. You are bothered by my writing "forty percent" without explaining what that means.

Hey. You are getting good at recognizing your mistakes.

Mea culpa.

Mea WHO?

I said *mea culpa.* That's Latin for "I fouled up."

So, what about the forty percent of the stores going out of business?

Forty percent means "forty out of a hundred." Forty percent can be written as 40%. For every 100 toy stores that treat their customers this badly, 40 of them go out of business within a month.

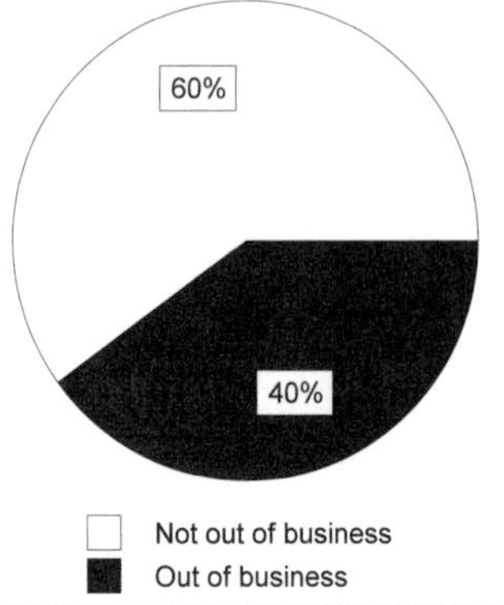

Pie charts are a nice way to graph quantities that add up to 100%. They are also called **circle graphs**.

40% means $\frac{40}{100}$ ✶

Your Turn to Play

1. KITTENS pays Fred a salary each month. Here is Fred's budget:

10% to Sunday school donation
50% to taxes
30% for books

How much is left over for other things (such as food and clothing)?

2. Continuing the previous problem, draw a circle graph of his budget.

3. Change $\frac{57}{100}$ to a percent.

4. Change $\frac{1}{4}$ to a percent.

5. Change 75% to a fraction. (Remember that one of the general rules is that you reduce fractions in your answer as much as possible.)

6. Change 0.31 to a percent.

7. $3\frac{1}{12} + 7\frac{1}{4} = ?$

8. Change 106.13% to a decimal.

Complete solutions are on page 195

✶ This reduces to $\frac{2}{5}$

The Bridge

from Chapters 1 – 20 to Chapter 21

first try

Goal: Get 9 or more right and you cross the bridge.

1. List the natural numbers between 30 and 40 inclusive that are composite. (Inclusive means including the extremes. In this case, the natural numbers between 30 and 40 inclusive are {30, 31, 32, 33, . . . , 39, 40}.)
2. Find two prime numbers that add to 40.*
3. The area of the circle is 49π square miles. What is the area of the square?

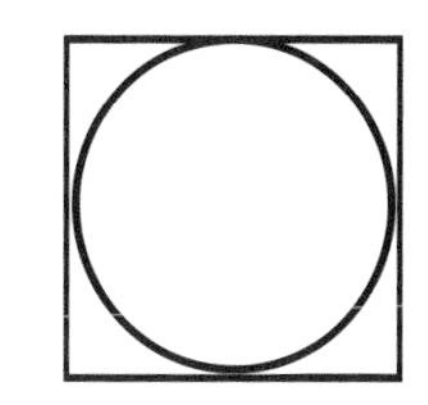

4. Change $\frac{1}{8}$ to a percent.
5. Fred spends 14% of his life reading; 31% teaching; 34% sleeping; 4% grooming and dressing; 5% at church; 11% talking with his friends. The rest of the time he spends eating. How much is that?
6. How many more square inches does a 12-inch pizza have than an 8-inch pizza? (A 12-inch pizza is a pizza with a diameter of 12". Use 3 for π in this problem.)
7. If a 12-inch pepperoni pizza costs $9.48, how much would one square inch cost? Round your answer to the nearest cent. (Use 3 for π in this problem.)
8. Change 7.003 to a mixed number. (Numbers like $9873\frac{1}{8}$ or $2\frac{1}{5}$ are called mixed numbers.)
9. $0.376 \div 0.047 = ?$
10. *Take a number, x, triple it, and then subtract 7* is a function. In algebra we will write this function as $f(x) = 3x - 7$. (It is read, "f of x is equal to 3x minus 7.") What is the inverse of this function?

✶ The Goldbach Conjecture states that there will always be *at least* one pair of prime numbers that add to any even number larger than 3. Sometimes there are more than one pair of primes that works. For example, 3 + 7 = 10 and 5 + 5 = 10.

The Bridge

from Chapters 1 – 20 to Chapter 21

second try

---

1. Find two prime numbers that add to 52.

2. Joe decided to try his luck fishing in a swimming pool. Darlene had made some ambrosia for him, and he had thrown it overboard. The circle represents the floating ambrosia. His boat is at the middle of the circle. What is the area of the the swimming pool *not* covered by the ambrosia?
(Use 3 for π in this problem.)

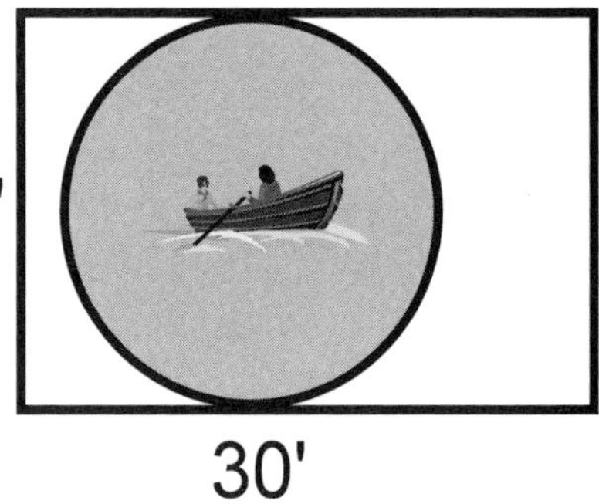

3. As everyone knows, ambrosia is made from oranges, shredded coconut, and sometimes pineapple. At one store, the oranges were \$0.37 per pound. At the second store they were 0.50¢ per pound. At which store were they cheaper?

4. The cost of the ambrosia was $\frac{1}{10}$ of Darlene's savings. What percent of her savings is that?

5. Darlene spent 2% of her savings on sun screen so she wouldn't get sunburned while she was sitting in the boat with Joe. Express 2% as a fraction. (And, as usual, reduce the fraction as much as possible.)

6. A shred of coconut costs 0.4¢. Darlene spent \$2.82 for the coconut for the ambrosia. How many shreds of coconut did she buy?

7. Darlene made 62.4 oz. of ambrosia for Joe. She used 39.7 oz. of oranges and 9.25 oz. of coconut. If the rest of the ambrosia was pineapple, how much pineapple did she use?

8. Joe started singing his favorite sea chanty. It had somewhere between 410 and 420 verses and the number of verses was divisible evenly by 9. How many verses did his sea chanty have?

9. The set of all people in Joe's boat is {Joe, Darlene}. Name a subset of this set.

10. $0.34 \times 34 = ?$

The Bridge
from Chapters 1 – 20 to Chapter 21

third try

---

1. What is the smallest number that is larger than a billion and is divisible by both 2 and 3?

2. Find two prime numbers that add to 54.

3. When Darlene was sewing her dress, she pictured a large circular ballroom where she and Joe would dance. The diameter of that ballroom was 32 feet. What was its area? (Use 3.14 for π in this problem.)

4. Round your answer to the previous problem to the nearest square foot.

5. Silver thread for her dress cost 3.4¢ per yard. A spool of silver thread contains 100 yards. How much would a spool cost? Give your answer in dollars.

6. The dress would cost 80% of Darlene's weekly salary. Express 80% as a fraction. (Please reduce the fraction as much as possible.)

7. Darlene estimates that when she and Joe are together, he is thinking about her one-twentieth of the time. Express that as a percent.

8. Darlene used 830 yards of gold thread for her dress. The thread costs $64.74. How much does a yard of gold thread cost?

9. If set A = { 🏇, ☎ }, then the subsets of A are { }, { 🏇 }, {☎} and { 🏇, ☎}.

How many subsets does B = {◘, ☼} have?

10. How many subsets does C = {x, y, z} have?
(Hint: the answer is even.)

The Bridge

from Chapters 1 – 20 to Chapter 21

fourth try

1. List the natural numbers between 40 and 50 inclusive that are composite.

2. One of the sailors who was visiting New York bought a bath mat with the Statue of Liberty on it. The diameter of the mat was 3 feet. He decided to sew some ribbon around the edge. How much ribbon would he need? (Use 3.14 for π.)

3. What is the area of that bath mat?

4. Round your answer to the previous problem to the nearest tenth of a square foot.

5. The bath mat cost fifty dollars. How much is that in cents?

6. The sailor realized that he couldn't use his purchase as a bath mat because that would mean that he would be standing on the Statue of Liberty. Instead, he decided to mail the 6.3-pound bath mat to his wife in Kansas. Because of rough handling, the mat lost 0.7 pounds. What was the weight of the mat that his wife received?

7. The mat lost 95% of its value because of the rough handling. Express 95% as a fraction. (Reduce the fraction as much as possible.)

8. Continuing the previous problem, draw a pie chart where the two sectors are "Value lost" and "Value remaining."

9. His wife used the bath mat as a liner for the kitty box. Because of that, the cat decided not to use the box three-eighths of the time. Express three-eights as a percent.

10. Find two prime numbers that add to 56.

fifth try

1. Find the area of the rectangle not covered by the two circles. (Use 3.14 for π.)

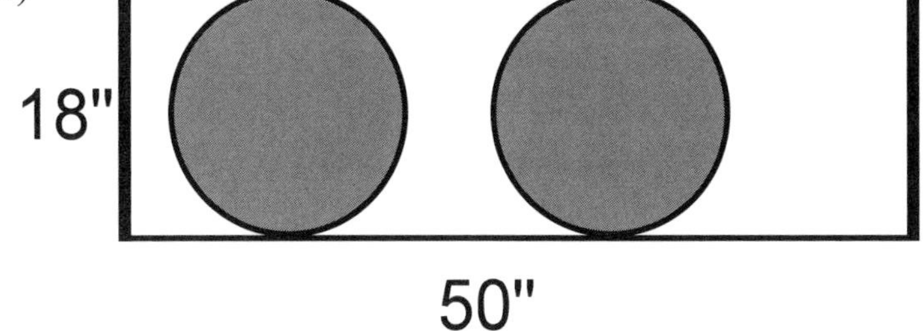

2. Write \$0.80 using the "¢" sign.

3. Which of these is largest: 6.5¢ \$0.09 7.00001¢

4. Change $\frac{7}{20}$ to a percent.

5. Change 68% to a fraction. (Reduce the fraction as much as possible.)

6. Bobbie, Berta, and Elizabot each weigh 477 grams. What is their combined weight?

7. Change 0.91 to a percent.

8. List the numbers between 60 and 70 inclusive that are prime.

9. Find three prime numbers that add to 21.

10. Change 43.203 to a mixed number. (Numbers like $9873\frac{1}{8}$ and $2\frac{1}{5}$ are called mixed numbers.)

Chapter Twenty-one
40% of 15

Forty percent of all toy stores that treat their customers badly go out of business within a month. When the cashiers tell the kids to "Stay in the center of the aisle and don't touch the toys," they violate the first rule of business which is: If you are going to sell to people, you must please them.

A **corollary*** is that if you don't please your customers, you are toast.**

Last month you could count 15 toy stores within walking distance of KITTENS university where you could hear the cashiers screaming at the kids. What's 40% of 15?

Here are two tricks you need to know:

Trick #1: "Of" often means multiply.

of = ×

Trick #2: This second trick is so obvious. Let me see if you can guess it before I tell you.

We know i) We have no idea how to multiply percents.
ii) We do know how to change percents into decimals.
iii) We know how to multiply decimals.

And so trick #2 is a corollary of these three facts that we already know. Are you ready? Here it is. [Drum roll please.] *When you are trying to figure out what 40% of 15 is, you change it to 0.4 × 15.*

---

* A corollary (CORE-ah-lair-ee) is a statement that comes as a consequence of what was just said, a natural result. We use corollaries a lot in geometry, because geometry is the one pre-college math course in which we prove many things. The things we prove are called **theorems**. Sometimes after we have worked long and hard to prove a theorem, there are one or more corollaries that follow easily.

Someone told me one Sunday morning that happiness is not the main theorem of life. It is not something you go after directly. Happiness comes as an easy result (a corollary) after you have done the hard work of "proving" the Main Theorem.

** *To be toast* is not some modern slang. It first came into our language in about 1375! That's long before they invented electric toasters. To be toast is to be in trouble or doomed.

Your Turn to Play

1. Last month there were 15 nasty toy stores. Forty percent of them are now out of business. How many went out of business?

2. Darlene made 62.4 ounces of ambrosia for Joe. 15% of that was coconut. How much coconut was in the ambrosia mix?

3. $7\frac{1}{12} - 3\frac{1}{4} = ?$

4. When Joe was singing the 414 verses of his sea chanty, he would often make up his own verses.

YO, HO, HO, AND A BOTTLE OF RUM. ♪

SIXTEEN RATS AND DUMB, DUMB, DUMB. ♫

In fact, he made up one-third of the verses. How many of the 414 verses did he make up?

5. Nine percent of the time Darlene is actively thinking about how her hair looks. What percent of the time is she not actively thinking about her hair?

Complete solutions are on page 196

Fred left the toy store and walked down the street. Behind him he could hear someone singing. He turned to look. It was Joe carrying a dripping boat. Joe was working on the 384th verse of his sea chanty.

YO, HE, HE, AND DO YOU SEE,
THE CAT AND THE CRADLE AND
THE DEEP BLUE FLEA. ♫ ♪

Darlene was walking several yards behind him. She was carrying the picnic basket and the fishing gear.

Fred called out, "Hi!" but Joe didn't hear him.

Darlene responded, "Hi, Professor Fred. It's been a long day." Joe kept on singing as he turned and headed into the boat rental store. Joe didn't notice the sign on the store window.

Darlene set her stuff on the sidewalk and sighed, "Sometimes that man. . . ."

Fred could guess how she might finish that sentence.

"I haven't had anything to eat all day today. Joe ate all the ambrosia that I made for our lunch," she said. "Do you know any good place to eat that is close by?"

Fred pointed. "It's a new place. It's called the 5100th Avenue Steak House. I have to warn you. They seem to serve very large portions."

"I'm so hungry, I could eat a cow," she said.

Joe joined them. "Hi Fred! When did you get here?"

Before Fred could respond, Darlene pulled on Joe's sleeve and said, "Come on, big boy. I hear there's a good steak house down the street."

"Been nice talking with you Professor Fred," Joe said as he turned and followed Darlene.

Fred was alone, again. He heard a tearing sound. The sign in the boat rental store window was being pulled down. Before Darlene and Joe were out of sight, a new sign was being taped up on the window.

That's quite a discount Fred thought. A rowboat normally rents for $24, and now the owner is offering a 30% discount.

There are two ways to find out how much the sale price is. There is the Hard Way and the Easy Way.

The Hard Way: First, find out how much you save.

$$30\% \text{ of } \$24 = 0.3 \times 24 = \$7.20$$

Then, subtract the savings to get the sale price.

$$\$24.00 - \$7.20 = \$16.80$$

The Easy Way: If it is 30% *off*, then it is 70% *on*.

$$70\% \text{ of } \$24 = 0.7 \times 24 = \$16.80$$

You can figure sale prices either way. The answer will be the same either way.

a small essay

"Why We Do Mathematics."

Mathematicians are just like human beings. They don't enjoy adding up columns of numbers or balancing their checkbooks. They don't multiply 398 by 975 for fun. They do math because it makes life easier. Doing math is the easiest way to make sure that our bridges don't fall down or our airplanes don't crash. Doing math is the easiest way to find out if you should buy that apartment house. (The hard way is to buy it because the bricks are cute, and then go broke.)

end of small essay

Your Turn to Play

1. A motorboat normally rents for $71. If you don't sing a sea chanty in the store, you get a 30% discount. How much will the rental price be after the discount?

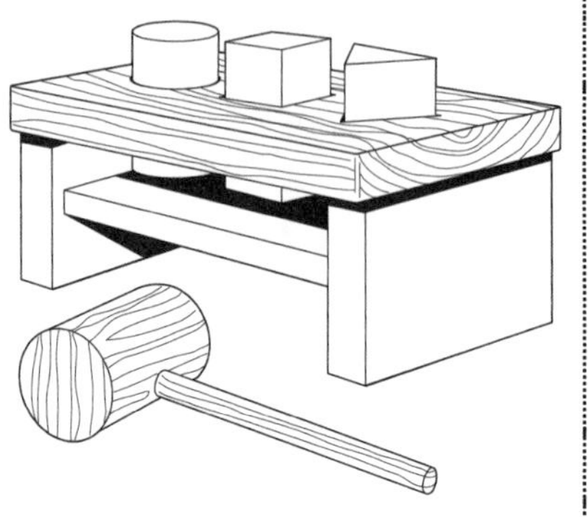

2. Just before the C.C. Coalback Toy Store closed, they had a going-out-of-business sale. Everything was one-third off the regular price. The regular price of the blocks and hammer was $41.31. How much was the sale price?

3. They were also offering a three-piece puzzle set. The original price was $20. It was on sale for one-third off that price.

You could get an additional discount of one-fourth (off the sale price) if you paid in cash. Finally, you could get another one-tenth off of the double-discounted price, if you agreed not to ask for a receipt. What would be the final price you paid?

4. Darlene went on a diet so that she would look nice for Joe. She started at 115 lbs. and lost 10% of her body weight. What was her new weight?

5. Joe invested $637.49 in the C.C. Coalback Investment Club. He lost 100% of his money. How much did he have left?

6. Fred had 84 students in his calculus class at the beginning of the semester. At the end of the semester he had lost 0% of them. How many students did he have left?

Complete solutions are on page 197

Chapter Twenty-three
Distance = Rate × Time

Having lived at KITTENS university for more than four years, Fred thought he knew all the local stores. Most of them didn't interest him at all. He seldom went to the grocery store because he bought most of his food at the vending machines down the hall from his office. (That's one reason why he is five years old and only three feet tall.) He goes to the bank to deposit his salary check and to the thrift store to buy his clothes.

But there is one thing that the town really lacked, and that was a bookstore. Fred rubbed his eyes. He couldn't believe what he was seeing. Just down the block was a flashing sign.

He started breathing fast. His pulse hit 119 beats per minute. He ran. One of his shoes fell off. He kept running.

He ran at the rate of 18 feet per second for 53 seconds. He stopped when he was right in front of the door. He had run 954 feet.

$$\frac{18\text{ feet}}{\text{sec}} \times \frac{53\text{ sec}}{1} = 954\text{ feet}$$

When you multiply rate times time you get distance. The formula in algebra is d = rt.

The "rt" means r × t.

It is easy to use d = rt. If you drive at 60 miles per hour for 3 hours, you will go 180 miles. If a snail goes at the rate of 0.1 inch per minute for 14 minutes, it will have traveled 1.4 inches. If Alexander is gaining weight at the rate of 3 pounds per year, in ten years he will be 30 pounds heavier.

Your Turn to Play

1. If Fred is saving his money at the rate of $1.47 per week, how much will he have saved in 52 weeks?

2. If Fred could run at the rate of one furlong* per minute, how far could he run in 2 hours? (Hint: We have mixed units here: hours and minutes. First, change the hours into minutes.)

3. When Fred started running, he weighed 37.5 lbs. When he lost his shoe, his weight decreased by 4%. How much did he weigh when he got to the door of the bookstore?

4. Change $\frac{1}{8}$ into a percent.

5. $7\frac{1}{12} \times 3\frac{1}{4} = ?$

Complete solutions are on page 197

✶ A furlong is an old-fashioned measure of distance. It is equal to 220 yards. Furlong was derived from the length of a furrow. (*furrow* + *long* = *furlong*)

Now you only have one question: What is a furrow? It is the long groove in the earth made by a plow. When you get worried, they sometimes say that you "furrow your brow."

Chapter Twenty-four
15% More

The sign on the door read, "No shoes, no service." Fred's little heart was beating fast—180 beats per minute. Did the sign mean that the bookstore didn't sell shoes and didn't offer any service? That seemed silly. As he settled down—and his heart rate dropped to 72—he realized that he was reading an elliptical* statement. Fred mentally rewrote the sign: BECAUSE OF FEDERAL LAW—AND OUR OWN SENSE OF MORALITY—WE DO NOT DISCRIMINATE ON THE BASIS OF SEX, RACE, HEIGHT, PLACE OF NATIONAL ORIGIN, OR MATH ABILITY. BUT YOU HAD BETTER BE WEARING SHOES—BOTH OF THEM!—OR WE WILL THROW YOU OUT.

That was clear enough. Fred walked back, retrieved his shoe, and put it on. That increased his weight by 4.17%.

Wait! I hate to stop you again, but I think you made another mistake, Mr. Author. Back in problem 3 in the previous *Your Turn to Play*, ***you said that his weight decreased by 4% when he lost his shoe. Now you tell me that his weight increased by 4.17% when he put his shoe back on. That sounds goofy.***

Goofy, but true. Let me give you a simple example. Suppose you ask people who have never read *Life of Fred: Decimals and Percents* the question, "If the value of your stocks goes down by 50% on Monday and goes up by 50% on Tuesday, are you back where you started?" Many of them will say yes, but they would be wrong.

Let's say that on Sunday your stocks were worth $200. If they lost 50% on Monday, they would then be worth $100. Now what does it mean to go up by 50%? Fifty percent of $100 is $50. On Tuesday, your stocks would be worth $150.

In Fred's case, he weighed 36 lbs. when he was wearing only one shoe. When he put on his second shoe, I said that his weight increased by 4.17%, and that he now weighs 37.5 lbs.

---

* When you say or write something and leave out lots of words, you are being elliptical.

As usual, there is the Hard Way and the Easy Way to do this problem.

The Hard Way: First, find out how many pounds he gained.

4.17% of 36 lbs. $= 0.0417 \times 36 = 1.5012$ lbs.

Second, add that 1.5012 to his starting weight.

$1.5012 + 36 = 37.5012$ which rounds to 37.5 lbs.

The Easy Way: A 4.17% gain means the original weight (100%) plus an extra 4.17%. $100\% + 4.17\% = 104.17\%$.

$36 \times 1.0417 = 37.5012 \doteq 37.5$ lbs.

Suppose that in the summertime it takes you 40 minutes to fire up your grill. It takes 15% longer in the winter. How long will it take?

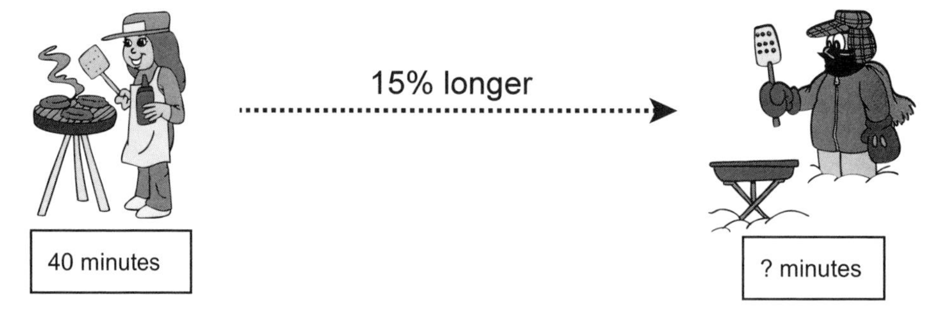

$40 \times 115\% = 40 \times 1.15 = 46$ minutes.

Your Turn to Play

1. Normally, it takes 7 lbs. of force to open the bookstore door, but because of the wind it takes 32% more force. How much force does it take with the wind blowing?

2. If an algebra book costs \$39, how much does it cost with an 8% sales tax added on?

3. Fred headed to the bargain book table. The sign on the table read: 3% OFF EVERYTHING ON THIS TABLE. Fred picked up a book by Prof. Eldwood entitled *Guide to Modern Ironing*, 1844. The regular price was \$87. What was the sale price?

4. Fred started reading the *Modern Ironing* book at the rate of 5 pages per minute. How many pages could he read in two hours?

5. Prof. Eldwood told his readers that if their irons are leaving wrinkles in their clothes, they should increase the weight of their 7-pound iron by 42%. How much would the iron weigh after the increase?

6. Prof. Eldwood told his readers that if the ironing takes too much effort, they should decrease the weight of their 7-pound iron by 42%. How much would the iron weigh after the decrease?

7. The handle on the 9-pound iron weighs 3 ounces. If you take the handle off, how much will the iron weigh?

8. He mentioned in the appendix to his book that there are three major brands of powdered starch: Star Brand, Bird Brand, and Drum Brand. He called this the starch set, S = {Star, Bird, Drum}. List all the subsets of S.

9. With a box of Bird Brand starch, you can do 56 shirts. With Drum Brand, you can do 25% more. How many shirts will Drum Brand do?

10. $7\frac{1}{12} \div 3\frac{1}{4} = ?$

Complete solutions are on page 197

Chapter Twenty-five
Area of a Triangle

Fred put down the *Modern Ironing* book and looked around. The store was huge. There was a section devoted to lawnmowers, and another devoted to life in Lithuania. *So much to look at!* Fred thought. *I'll be here till they close and they kick me out.*

The different sections of the store were each in different shapes. The kids' section of books was arranged in a circle. The store owners called it the Children's Circle.

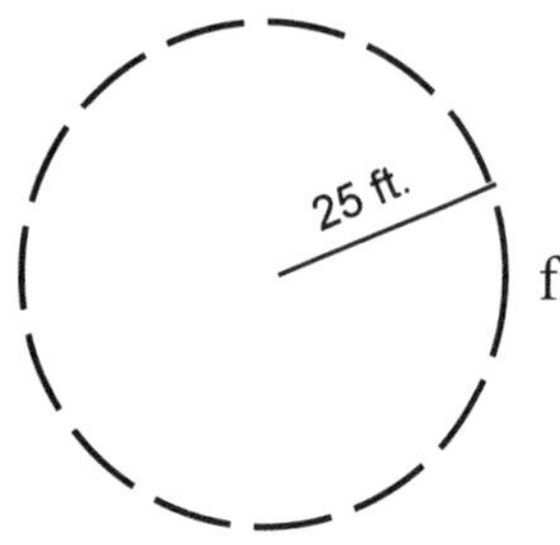

Using $A = \pi r^2$ and approximating π by 3, Fred figured in his head: $A = 3 \times 25 \times 25 = 1875$ square feet.

The lawnmower section of the bookstore looked like a green lawn. (If this is your book, feel free to color the rectangle green. Many of today's math books have lots of color printing in them. That has two drawbacks: ① You lose the fun of coloring, and ② it makes the books co$t a lot more.) In the lawnmower section, there were books like *Loving Your Mower* and Prof. Eldwood's *Why Your Work is Not Done When the Cows Mow Your Lawn*, 1855.

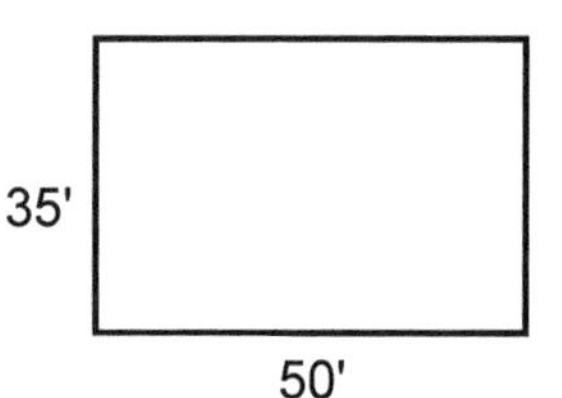

The area of a rectangle is equal to length times width. $A = lw$ which in this case is equal to 35×50.

Note to my readers: Now I am going to ask you to do something new.

New! Always something new! This whole book has been full of new stuff. Why can't we just have the old and familiar—like my teddy bear—so I don't have to strain my brain?

Almost by definition, learning things means doing new stuff. If you always do the things you already know, you don't get any smarter. And—you are right—doing new stuff does involve what you call "brain strain."

You have a choice:

I'll give it a try

No brain strain for me

Okay. What new stuff did you want me to try?

We have the area of the rectangle equal to 35 feet times 50 feet. What I want you to do is multiply these two numbers together *in your head.* No paper. No pencil. Just thinking. It may take a minute or two. It's your choice: Make the effort or sit on the bench.

I will wait.*wait wait.*

If your answer came out 125 square feet less than the Children's Circle area of 1875 square feet, you did it right.

Another popular section of the bookstore featured books on teepees. The best seller in that section was Prof. Eldwood's *Financing Your Teepee with an Adjustable Rate Mortgage*, 1853. This section of the store looked like a teepee. It was triangular.

From the cover of Prof. Eldwood's book

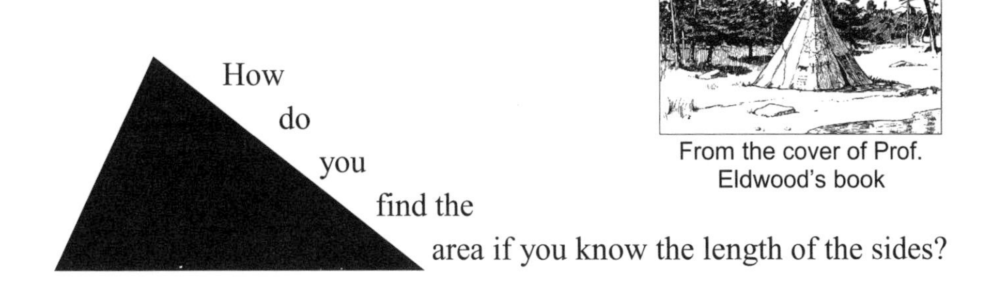
How do you find the area if you know the length of the sides?

Now suppose you know the lengths of the three sides. For example, suppose a = 9', b = 12', and c = 15'.

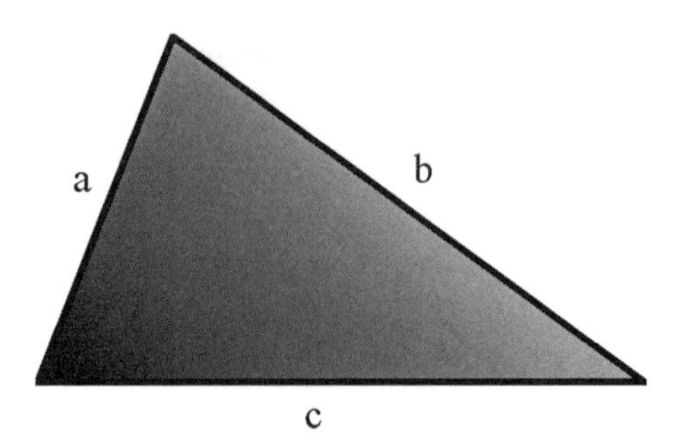

Question: How do you find the area?
Answer: You can't.

You could, if you knew a little algebra. But not right now.

Hey! Forget that nonsense about having to wait! I want it now. Now! Do you hear me?

But this is an arithmetic book. No author ever presents **Heron's formula** at this level. It would drive the readers crazy.

I'm already crazy. Give me Hero's formula—or else!

It's Heron, not Hero. Heron of Alexandria was an engineer who lived about. . . .

Yeah. I know what a heron is. It's a long-necked, long-legged wading bird. I'll even draw you a picture of a heron.

heron

Thanks. That's a pretty picture. But the heron bird is pronounced HAIR-en. But my guy, Heron, is pronounced HAIR-on. I would like to put a photograph of Heron in this book, but I couldn't find one since Heron lived either at the same time as Christ or within about forty years after his death.

Heron invented the first recorded steam engine, but it was thought of as just a toy. It was centuries later that steam engines for locomotives were created.

He also was the inventor of the first vending machine—drop in a coin and it would dispense water.

Hey! You're getting off the track. I want to see Heron's formula, so that when I know the three sides of a triangle, I can compute its area.

I didn't get "off the track." I'm just scared. I'm afraid you'll hate me if I give you the formula.

Give it to me.

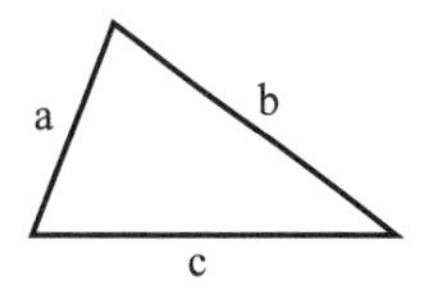

Okay. Start with a triangle where you know the lengths of the three sides: a, b, and c.

Next, compute the perimeter (which is the distance around the outside). $p = a + b + c$.

You haven't lost me yet.

Then compute the semiperimeter (which is equal to half of the perimeter). $s = \frac{p}{2}$ or $s = (\frac{1}{2})p$ or $s = p/2$.

So far this is super easy. On the previous page you wrote, "suppose a = 9', b = 12', and c = 15'." ***I can find the perimeter. It's p = 9 + 12 + 15, which is 36. And finding the semiperimeter is duck soup. s = 18. So what is so hard?***

The next step: According to Heron's formula, the area of a triangle whose sides are a, b, and c is:

$$A = \sqrt{s(s-a)(s-b)(s-c)}$$

Wow. That looks like a killer.

I warned you. Nobody ever writes that in an arithmetic book.

You did.

You made me.

Some of that I can do. I can put in the values for s, a, b, and c:

$$A = \sqrt{18(18-9)(18-12)(18-15)}$$

and I can do the subtraction:

$$A = \sqrt{18(9)(6)(3)}$$

18(9)(6)(3) means 18 × 9 × 6 × 3

and I can do the multiplying:

$$A = \sqrt{2916}$$

but I have NO IDEA what that $\sqrt{\quad}$ ***thing means.***

$\sqrt{\quad}$ stands for *take the square root of.* Earlier in this book we talked about functions such as *add 3.* Then we defined inverse functions. The inverse function of *add 3* is *subtract 3.* The inverse function undoes what the function did.

In Chapter 7 we looked at the function *square the number*. The square of 5 is 25. The square of 9 is 81.

The inverse function of *square the number* is *take the square root of the number.* The square of 5 is 25. The square root of 25 is 5.

So $\sqrt{25} = 5$. $\sqrt{100} = 10$. $\sqrt{9} = 3$.

So all I have to do to finish the problem is find the square root of 2916. What number times itself equals 2916? By trial and error . . .

20 × 20 = 400 Too small.

90 × 90 = 8100 Too large. I want it to equal 2916.

50 × 50 = 2500 Now we're getting closer.

55 × 55 = 3025 Too large

54 × 54 = 2916 I got it.

So A = $\sqrt{2916}$ = 54 square feet is the answer.

Would you like to see an easier way to find the area of a triangle?

Yes, please.

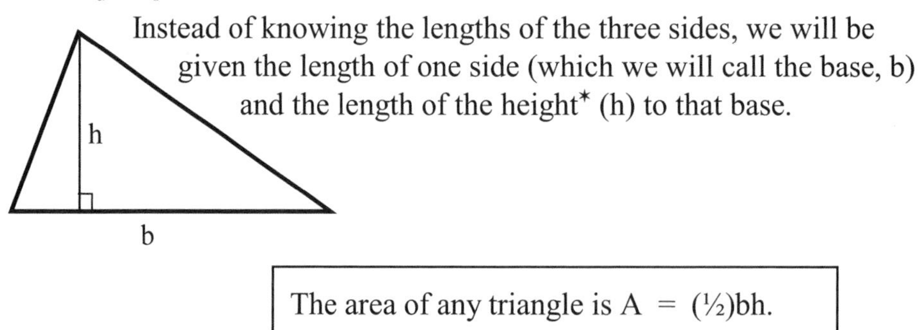

Instead of knowing the lengths of the three sides, we will be given the length of one side (which we will call the base, b) and the length of the height* (h) to that base.

The area of any triangle is A = (½)bh.

So if the base were b = 15, and the height were h = 7.2, then the area of the triangle would be A = (½)bh = (½) × 15 × 7.2 = 54.

✶ In geometry the height is called the altitude. The altitude makes a 90° angle with the base. A 90° angle is called a right angle. The standard way to indicate a right angle is by drawing a little box where the two lines meet.

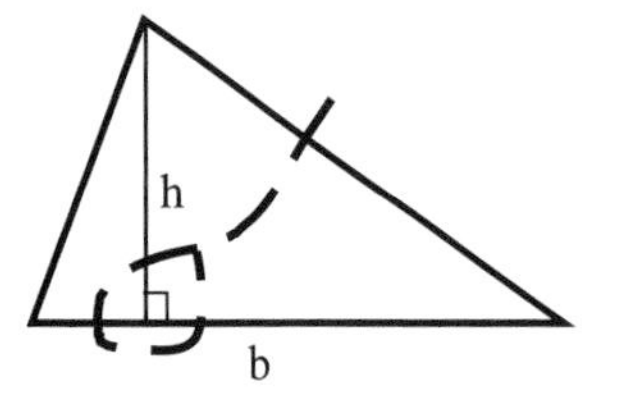

Your Turn to Play

1. Here is one problem to do in your head. Please give it a try. Find the area of a rectangle that has a width of 22 feet and a length of 33 feet. Figure out the answer before you look at the answer given below.

2. Find the value of $\sqrt{5184}$. You will probably have to make several guesses before you find the number that when squared will equal 5184.

3. Here is a triangle in which the altitude (the height) falls outside of the base. That is okay. The formula still works. Find the area of the triangle.

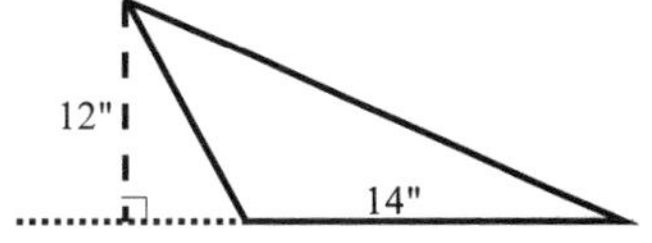

4. Find the area of a triangle with a base of 7.8' and a height of 6.9'.

5. When your aunt died, she left you 27 square miles of land. It was all the land between three towns. The distance from Appleton to the Watermelon Freeway was 9 miles. How far is it from Beeville to Seaside?

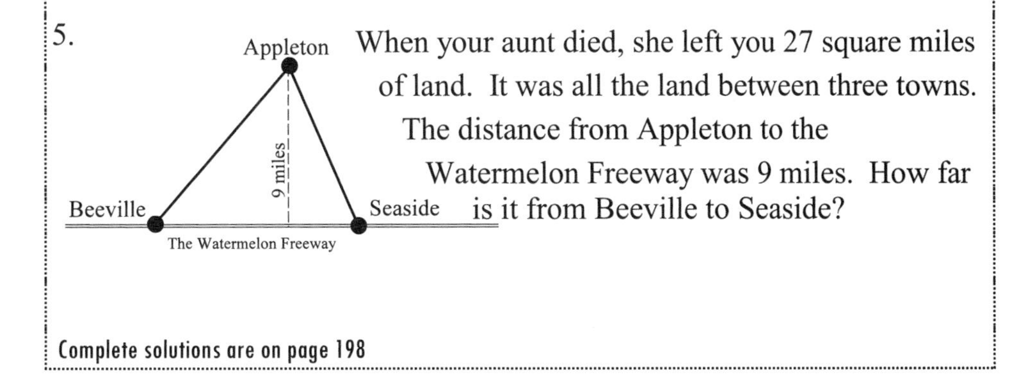

Complete solutions are on page 198

The Bridge

from Chapters 1 – 25 to Chapter 26

first try

Goal: Get 9 or more right and you cross the bridge.

1. Fred spends 14% of every 24 hours reading. How many hours is that?

2. Fred read 51 books last month. Two-thirds of them were nonfiction. How many nonfiction books did he read?

3. Fred has 2100 books in his office. He lent 37% of them to students. How many books are still in his office?

4. Fred has his office in the Math Building. The building is one furlong (220 yards) long. Fred's office is 4.5% of the length of the building. How many yards long is his office?

5. After teaching his math classes, there is nothing that delighted Fred more than to curl up in his office and read. He reads math (of course), and history, and Shakespeare, and the Bible, and poetry. One of his favorite poets is Christina Rossetti.

In the wintertime, when the snow piled up outside his office window, Fred liked to read Rossetti's "Winter: My Secret," which begins

I tell my secret? No indeed; not:
Perhaps some day, who knows?
But not today; it froze, and blows, and snows,
And you're too curious: fie!
You want to hear it? well:
Only, my secret's mine, and I won't tell.

Fred would read her poems at the rate of 6 poems per hour in the winter. (Her poems weren't that long, but he liked to stop and think about what he had read.) How many poems could Fred read in 3.5 hours?

6. In the summer he would read her poems 20% faster. What was his summer rate of reading Rossetti's poems?

7. By trial and error, find $\sqrt{1521}$.

8. What is the area of a triangle with base = 6" and altitude = 7.7"?

9. $8.7 - 3.07 = ?$

10. Express 7° in minutes.

The Bridge

from Chapters 1 – 25 to Chapter 26

second try

1. Darlene has \$333.50 in savings. She spent 2% of that on sun screen that she uses when she goes fishing with Joe. How much did the sun screen cost?

2. Joe saw on television that his favorite fishing lure, which normally costs \$14, was on sale for 7% off. What was the sale price of the lure?

3. Ninety-seven percent of the verses in Joe's sea chanties have the phrase *yo, ho, ho* in them. What percent don't have that phrase?

4. The wind blew Joe's boat across the swimming pool at the rate of 6 inches per hour. How far did it move in 7.3 hours?

5. Convert your answer to the previous problem into feet and round it to the nearest foot.

6. Joe figured that his new lure would increase the number of fish he caught by 30%. Before he bought the lure, he caught 0.67 fish per hour. How many fish per hour did he estimate he would catch with the new lure?

7. When Joe and Darlene were out fishing in the swimming pool, Joe told her that he was "bothered by all the corners that the swimming pool had." Darlene couldn't understand him. She explained to him that all rectangular swimming pools have four corners.

"That's too many," he responded. "The pool I want has three corners." He drew a diagram. "And it will be 15 feet from the diving board to the opposite side." What is the area of Joe's pool?

8. Find two prime numbers that add to 66.

9. What is the area of the largest possible circle that will fit inside a rectangle that is 9" by 11"? (Use 3 for π.)

10. By trial and error, find $\sqrt{6561}$.

The Bridge

from Chapters 1 – 25 to Chapter 26

third try

1. Darlene could get 69.3 yards of thread from a large sheep. She could get 70% of that from a small sheep. How many yards of thread could she get from a small sheep? Round your answer to the nearest tenth of a yard.
2. Darlene bought 640 sequins to sew on her dress. Ninety percent of them were purple. How many were purple? (Sequins are tiny shiny disks.)
3. When Darlene's spinning wheel was new, she could spin it at 88 revolutions per minute. She once took it on a picnic with Joe. It rained and her spinning wheel got a little rusty. She can now spin it at only 67% of her original rate. How fast can she spin it now? Round your answer to the nearest revolution per minute.

4. Darlene could spin thread off of her spinning wheel at the rate of 19 inches per minute. How much thread could she spin in an hour and a half?
5. Convert your answer to the previous problem into feet and round it to the nearest foot.
6. When Darlene was sewing sequins onto her dress, she did 2.7 per minute. Since she had 640 sequins to put on her dress, she wanted to speed up the process. She got a glue gun and found she could glue them on the dress 71% faster than sewing them on. How fast was she putting them on with the glue gun?
7. She cut out one piece of satin for a skirt panel. → What was its area?

13.3"
27"
skirt panel

8. There are only two consecutive numbers that are both prime. Name them.
9. Name the smallest whole number that is evenly divisible by 98 and 5555.
10. By trial and error, find $\sqrt{4356}$.

The Bridge
from Chapters 1 – 25 to Chapter 26

fourth try

---

1. When the sailor mailed the 6.3-pound bath mat home to his wife in Kansas, it lost 11% of its weight because of rough handling. How much weight did it lose? Round your answer to the nearest hundredth of a pound.

2. If the bath mat lost 11% of its weight, what percent did it retain?

3. The bath mat normally sells for \$69.44, but the sailor bought it at a 28% discount. How much did he pay? Round your answer off to the nearest cent.

4. When the sailor's wife received the bath mat, she showed it to all of her friends. They had never seen a bath mat with the Statue of Liberty on it and were very envious. Because it is not very wise to have envious friends, she decided to rent the bath mat to her friends. She charged \$1.57 per hour. She rented it for 3 hours. How much rent did she receive?

5. She found that if she put a little perfume on the bath mat, she could charge 8% more than her usual rent of \$1.57 per hour. What is the new rental rate? Round your answer to the nearest cent.

6. Five-eighths of her friends decided to go into the bath-mat renting business. Express five-eighths as a percent.

7. The diameter of the bath mat was 3 feet. Using 3 for π, find its area.

8. What is the area of the largest possible circle that will fit inside a rectangle that is 60 miles by 90 miles? (Use 3.14 for π.)

9. Find two prime numbers that add to 30.

10. By trial and error, find $\sqrt{3249}$.

The Bridge

from Chapters 1 – 25 to Chapter 26

fifth try

1. When Fred teaches, he uses 8 sticks of chalk every hour. How many sticks would he use in a 90-minute class?
2. Three-fourths of the 3868 students at KITTENS University are enrolled in Fred's math courses. (The other fourth have already taken every math course Fred teaches.) How many students are enrolled in Fred's math courses?
3. The algebra textbook that KITTENS required before Fred started teaching was Prof. Eldwood's *Algebra and Algebra Revisited*, 11th edition. It cost $97.50. Fred saved 60% for his students when he switched to *Life of Fred: Beginning Algebra Expanded Edition.* How much is the *Life of Fred: Beginning Algebra Expanded Edition* book?
4. Change $\frac{1}{2}$ into a percent.
5. The 11th edition of Prof. Eldwood's *Algebra and Algebra Revisited* cost $97.50. The 12th edition cost 30% more. How much did the new edition cost?
6. By trial and error, find $\sqrt{1156}$.
7. What is 87% of 14?
8. Find the area of the circle. (Use 3.14 for π.)

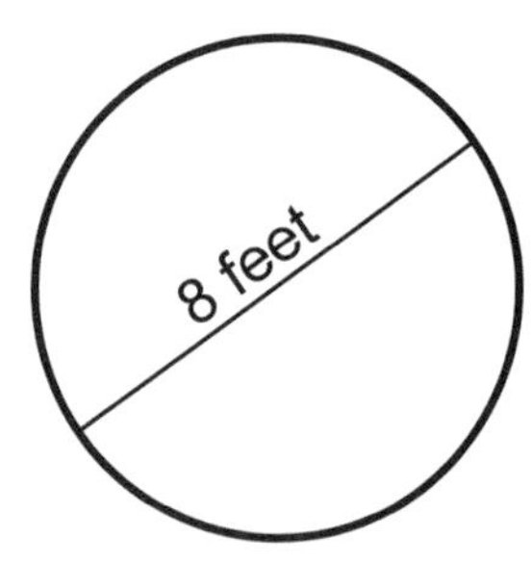

9. $8.08 + 17 + 9.9 = ?$

10. Find the area of the triangle.

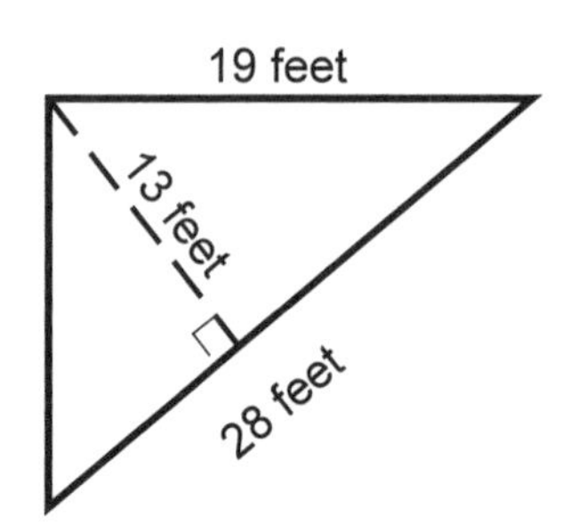

Chapter Twenty-six
Area of a Parallelogram

Fred rushed around from section to section in the bookstore. He didn't spend much time in the lawnmower section because he was too short to push a lawnmower. He didn't even stop at the Children's Circle. Those books were so—he had trouble thinking of the right word—they were so immature.

The section devoted to Franklin Pierce was in the shape of a parallelogram.* Fred stopped in this section to pay his respect to the former president. He opened Prof. Eldwood's *The President Who Came Between the 13th and the 15th Presidents,* 1858, and read that Pierce's nickname was Young Hickory of the Granite Hills.** That's a strange nickname Fred thought. It's longer than his regular name. I would have called him Frankie.

Just then Fred felt the back of his shirt tighten. Someone had seized*** him by the back of his shirt and had lifted him off the ground.

"Hey kid. Did you get lost?" the bookstore guard asked. Without waiting for an answer, the guard carried Fred over to the Children's Circle. Fred felt like a kitten being carried by a mother cat. Fred wanted to explain to the guard that he was a professor of mathematics at KITTENS, but the guard didn't give him a chance.

✶ Parallelograms look like this: ▱ They have four sides, and the opposite sides are parallel. Squares are parallelograms, and so are rectangles.

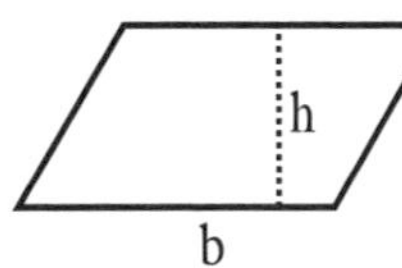

The area is equal to base × height. $A = b \times h$

✶✶ That's not made up. Young Hickory of the Granite Hills is his official nickname. If you are ever on a television quiz show, that may come in handy.

✶✶✶ English is much harder than mathematics.

Did you ever hear of the rule: *i before e, except after c*? The word *seized* doesn't seem to fit that rule. Neither does the word *neither.* Nor does the word *science.*

The guard lifted Fred over the top of the metal fence and dropped him into the Children's Circle.

Stunned, Fred looked around. It was a cage! This looks like a kids' concentration camp Fred thought to himself. On the chainlink fence was a sign:

I'm doomed! I don't have a mother or father to pick me up. I'll be here forever.* Being trapped in a bookstore forever wouldn't be bad, but I'm stuck in the kids' section. Fred picked up a book that was lying on the floor. It was entitled *See Dick Run.* He opened it and read:

See Dick run.
Run Dick run!
See him run, run, run.
What fun. See him run.

Fred wanted to throw up. This would be like being forced to watch daytime television. Fred was afraid that his mind would rot.

What to do?

He put the *See Dick Run* book on the shelf and looked around. He had never seen so many little kids running around unsupervised. And look! There are little tables and chairs that are just my size Fred thought. Fred ran over and sat down at the table. It felt comfortable. For more than four years, Fred had sat on three phone books, so that he could be tall enough to use the adult-size chair and desk in his office.

Children's books, papers, and broken crayons were scattered all over the table. He put the broken crayons in a crayon basket. He stacked the books in a neat pile.

---

* As you may have noticed, five-year-old Fred is parentless. When you get to *Life of Fred: Calculus Expanded Edition*, you can read the story of what happened when he was six months old.

The top book on the pile was entitled *See Jane Jog: Second in the Series.* Had Fred opened it—which he didn't—he would have read:

See Jane jog.
Jog Jane jog!
See her jog, jog, jog.
What fun. See her jog.

He took one of the pieces of paper that didn't have too much scribbling on it and drew a parallelogram.

He knew that the area was A = bh.

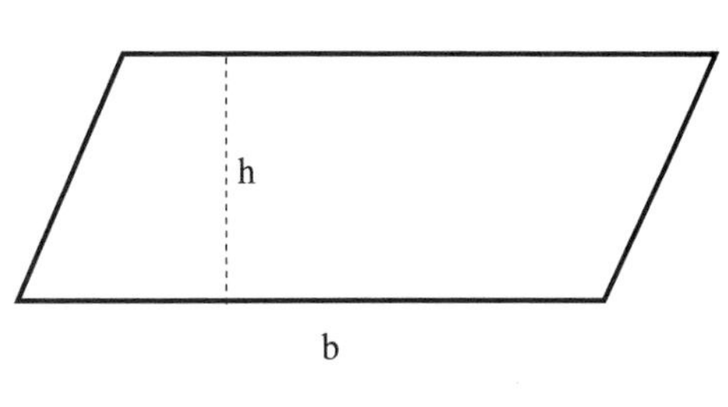

Then he took scissors and cut off a triangle from the right side.

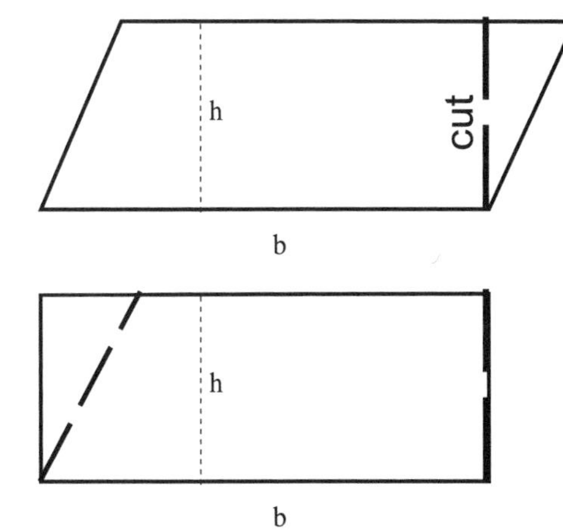

Then he pasted it to the left side. It turned into a rectangle. The area of the rectangle is equal to the area of the original parallelogram.

The area of the original parallelogram was A = bh. So the area of the rectangle also has to be equal to bh.

But b is the length of the rectangle, and h is the width of the rectangle. And we already know that the area of a rectangle is length times width.

It all fit together so nicely.

He tried the scissors on the chainlink fence to see if he could cut a hole and escape. It didn't work.

He drew another parallelogram and cut it along one of the diagonals. The two triangles were congruent.* And each of the triangles had an area equal to (½)bh. Again, everything fit together nicely.

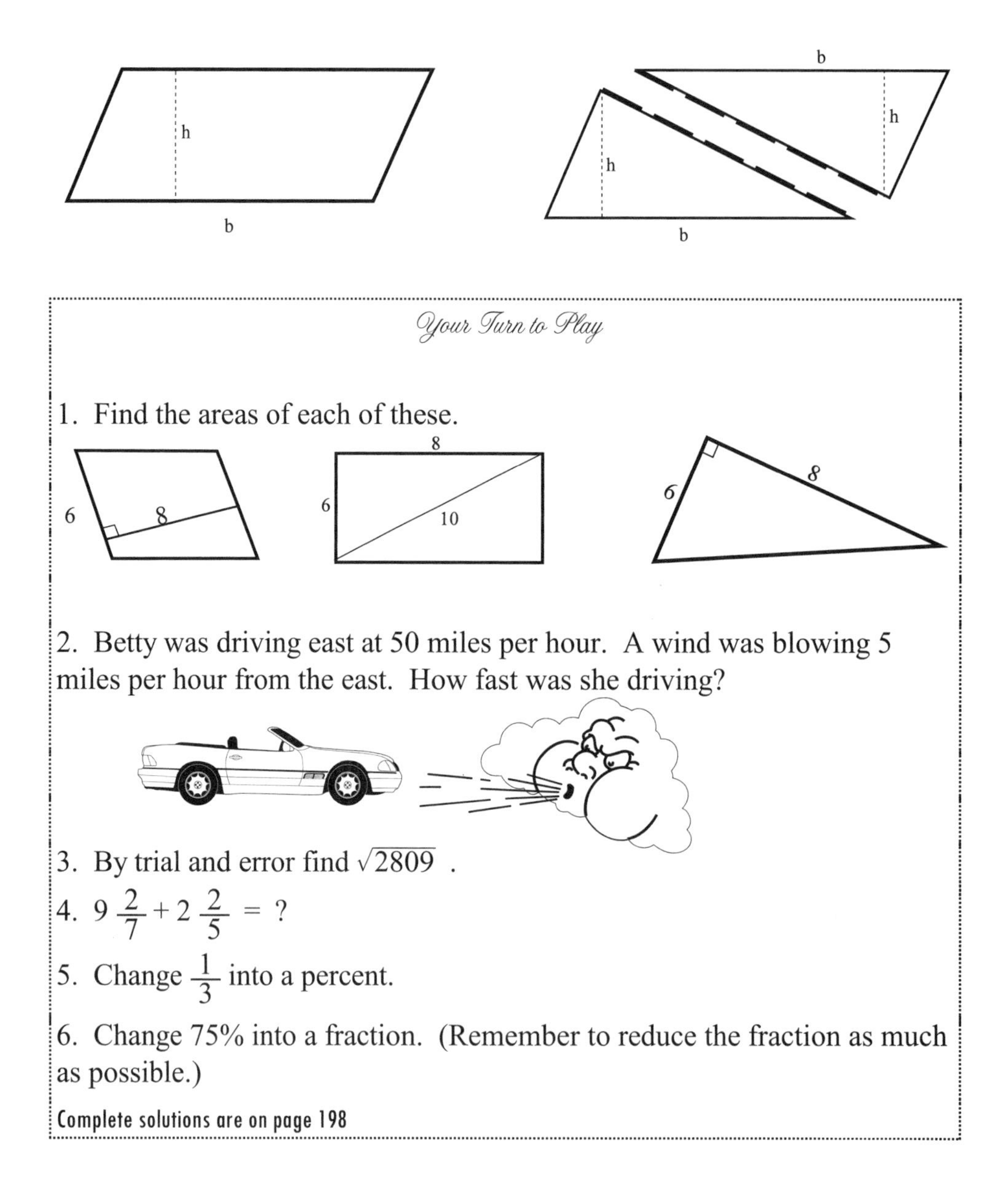

Your Turn to Play

1. Find the areas of each of these.

2. Betty was driving east at 50 miles per hour. A wind was blowing 5 miles per hour from the east. How fast was she driving?

3. By trial and error find $\sqrt{2809}$.

4. $9\frac{2}{7} + 2\frac{2}{5} = ?$

5. Change $\frac{1}{3}$ into a percent.

6. Change 75% into a fraction. (Remember to reduce the fraction as much as possible.)

Complete solutions are on page 198

✶ (con-GREW-ent) Two triangles are congruent if they are the same size and shape.

Chapter Twenty-seven
13 Is What Percent of 52

A three-year-old noticed that Fred had stacked all the books in a neat pile. He thought that Fred was a "libearian." He took a book off the shelf, handed it to Fred, and said, "Read me."

The book was about a cat named Puff. Fred opened it and read:

See Puff run.
Run Puff run!
See him run. Puff! Puff!
What fun. See him run.

Without saying thank you, the little boy wandered off to play in another part of the cage. Fred put the book on top of the pile of books he had made.

I have got to get out of here. There has to be a gate to this cage. He walked around the perimeter of the cage looking for an opening. (A couple of chapters ago we noticed that the Children's Circle looked like this: The radius of the circle was 25 feet. The diameter would then be 50 feet, and so the perimeter of the circle (the circumference) would be 50π. That is about 150 feet.)

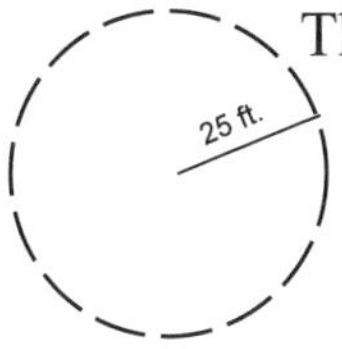

25 ft.

Fred found the gate, but it could only be opened from the outside.

He thought of digging under the fence. He had seen in movies prisoners escape by digging under fences. He stomped his foot. Under the carpet, he could feel concrete. That would be pretty tough digging.

He imagined himself growing old in the Children's Circle. Year after year would pass, and he could never get out. He would always stay three feet tall because he had nothing to eat. When he got to be a teenager, he would start growing a mustache. By the time he was forty, he would have a full beard. Ten years after that, he would start to get gray hairs in his beard. And still, no one would come to rescue him. Then old age. He would have spent most of his life reading children's books to kids who would come up to him and say, "Read me." Then one day he would die and pass on to the life hereafter. There he would see God. And he would be thanked by the thousands of adults who had fond memories of Fred reading *Run* books to them when they were children.

Fred felt his chin to see if whiskers were starting to grow.

An hour later, a mother came to pick up her three-year-old.

I know what to do! Fred thought. He walked up to the woman and whispered, "Will you adopt me?"

She ignored him and left with her kid who pointed to Fred and said, "Libearian," which is how some little kids say *librarian*.

I could climb the fence! In the movie West Side Story I saw guys climbing over tall chainlink fences. Fred tried and slipped.

The owners of Books! Books! Books! Bookstore had anticipated that some children would try to climb out. They put some grease on the fence so that it was too slippery to climb.

The fence was 52 inches high. Fred looked around to see if there was any way to get over the fence without climbing it. The table! Fred tried to pull on the table, but he found that it was too heavy to move.*

The books! I can move those. Fred moved his pile of books over to the fence a few at a time. It made a stack that was 13 inches tall. I'm 25% done!

Wait! Stop! You did it again. I, your reader, don't understand how you got that 25%. Did Fred just make up that number?

Fred doesn't make up numbers. You know how to do the calculation: 25% of 52 inches equals 0.25 times 52, which works out to 13 inches. Six chapters ago, we pointed out that *of* often means multiply. So when I want to solve "25% of 52 = ?" I just multiply together 25% and 52.

But. But. But. You weren't given the 25% and the 52. You started out with the problem "What percent of 52 is 13?"

---

* Actually, the bookstore owners had anchored it to the floor with screws so that it couldn't be moved.

It looks like ?% × 52 = 13. We have never done that kind before.

Oops. You are right. We have only done "25% × 52 = ?".

Perhaps, you would care to explain how to do this new kind of problem?

I guess that's my job. Okay. If you know both sides of the *of*, then you multiply. 25% of 52 means multiply 25% by 52.

I know that!

Patience please. I was just getting warmed up. As I was saying, if you know both sides of the *of* you multiply. If you don't, you divide.

Divide what into what?

You divide the number closest to the *of* into the other number. We started with the question "13 is what percent of 52?" That could be written as 13 = ?% of 52. The number closest to the *of* is the 52. You divide 52 into 13.

$$52\overline{)13.00} = 0.25 \quad \text{and } 0.25 = 25\%$$

Here is a ton of examples:

What is 6% of 39? ➔➔➔➔➔➔➔ 0.06 × 39

7 is what percent of 40? ➔➔➔➔➔➔➔ $40\overline{)7.00}$

82 is 2% of what? ➔➔➔➔➔➔➔ $0.02\overline{)82.00}$

91% of what number is 55? ➔➔➔➔➔➔➔ $0.91\overline{)55.00}$

What percent of 397 is 6? ➔➔➔➔➔➔➔ $397\overline{)6.00}$

This is the hardest part of percents. I will put it in a box so it will be easy to find.

> If you know both sides of the *of*, then multiply.
> Otherwise, divide the number closest
> to the *of* into the other number.

Your Turn to Play

1. The fence is 52" tall. Fred estimated that he would need a stack of books 75% of the height of the fence in order to climb over the fence. How high would that stack be?

2. In the previous *Your Turn to Play* we found that 75% $= \frac{3}{4}$

Do the previous problem using fractions instead of decimals

.

3. If Fred's stack is 26" tall, what percent of the height of the fence is it?

4. If too many kids were escaping over the 52" fence, the Books! Books! Books! Bookstore owners might decide to make the fence 40% taller. Then how tall would it be?

5. Convert your answer to the previous problem into feet and inches.

6. A chainlink fence that was a little higher than six feet might have some drawbacks. The guard had lifted Fred over the fence and put him in the Children's Circle. If the fence had been higher than six feet, the guard might have thrown Fred over the fence. That would not have been so nice.

The three owners had a meeting, and two of the three voted to keep the fence at 52 inches. What percent voted to keep the height of the fence?

7. $9\frac{2}{7} - 2\frac{3}{5} = ?$

Complete solutions are on page 199

Chapter Twenty-eight
Ratio

Fred continued piling books up next to the fence until he had a stack that was as high as the fence. There was only one problem. He couldn't climb on top of the stack. So he made a second stack that was 39 inches tall. And a third stack that was 26 inches tall. And a fourth stack that was 13 inches tall. He called it his Stairs to Freedom.

It took him twenty minutes of hard work to build his stairs, and he hadn't noticed what was happening outside the Children's Circle. A crowd had gathered to watch Fred at work. When Fred climbed the stairs, he was met by a sea of people.*

Some of the people began applauding. They had never before seen a little kid figure out how to escape from the Children's Circle. But one man had his hands on his hips. He was the bookstore guard. "Just what do you think you're doing?" he asked rhetorically.**

"It's all right, officer," a woman said. "He's with me."

Fred couldn't believe his eyes. It was Betty! Fred jumped into her arms (and nearly knocked her over).

Betty carried him to another part of the store to get away from the crowd and the guard. Fred whispered in her ear how he had lost his shoe, looked at a book on ironing, visited the lawnmower section, read about Franklin Pierce, etc. He was talking a mile a minute, and Betty couldn't understand what he was saying.

✶ Calling the crowd of people a *sea* is a metaphor. A metaphor (MET-a-four) is a figure of speech. Metaphors make a comparison boldly. They don't use words *like* or *as*. (If I had written that the people were *like* a sea, it wouldn't be a metaphor.) We use tons of metaphors in everyday speech. ("Ton" is used metaphorically in the previous sentence.) We call the vertical supports for a table *legs*.

✶✶ A rhetorical (re-TORE-a-cul) question is a question that isn't asking anything. The guard knew exactly what Fred was doing. He wasn't expecting Fred to give an answer like, "I believe it is obvious to everyone that I am attempting to escape." Such an answer would have made the guard angry. Silence is often the best answer to rhetorical questions.

Fred had managed to utter 268 words in 67 seconds. The ratio (RAY-she-oh) of 268 words to 67 seconds is $\frac{268 \text{ words}}{67 \text{ seconds}}$ which works out to $67\overline{)268}$ with quotient 4 $= \frac{4 \text{ words}}{\text{second}}$

". . . and then you came," Fred finished up. He had told the entire story in a single breath.

"Look! A book signing," said Betty. "I wonder who the author is." Betty set Fred down and took his hand. They headed over to the table where the author was signing his books. Hidden behind large stacks of books was Prof. Eldwood. He was signing his newest book, *Modern Clown Masks*, 1862. Fred thought that he looked very old.

Betty bought a copy and had the author sign it. The pages were yellowed with age. It had been printed before the advent of acid-free paper.

Next to the books were some clown masks for sale.

"How much are the masks?" Betty asked.

Prof. Eldwood thought for a moment. "I can let you have them . . . let's see . . . how about three masks for \$54."

"I would like all five please."

Prof. Eldwood scratched his head. "Let's see. If \$54 buys 3 masks, then the conversion factor is $\frac{\$54}{3 \text{ masks}}$ since \$54 equals 3 masks.

Using the conversion factor, $\frac{5 \cancel{\text{masks}}}{1} \times \frac{\$54}{3 \cancel{\text{masks}}} = \frac{\$270}{3} = \$90$."

Your Turn to Play

1. Fred spotted some leather bookmarks for sale. I can always use bookmarks he thought to himself. They were priced at \$8.61 for 7 of them. How much would 11 of them cost?

2. \$8.61 seemed very expensive for 7 leather bookmarks. Fred noticed a sign that read OPENING DAY SPECIAL FOR BOOKMARKS—5% OFF THE REGULAR PRICE. To the nearest cent, how much would 7 bookmarks cost with this 5% savings?

3. Fred wished that the sign had read OPENING DAY SPECIAL FOR BOOKMARKS—ONE-THIRD OFF THE REGULAR PRICE. How much would 7 bookmarks cost with this savings?

4. There were 8 women and 6 men standing at Prof. Eldwood's table. What is the ratio of women to men?

5. $9\frac{2}{7} \times 2\frac{3}{5} = ?$

Complete solutions are on page 200

Chapter Twenty-nine
Ordered Pairs

Fred was always thinking about ways to make his classroom teaching more lively. For Fred, the most important thing was not to be boring. Students never fell asleep in his classes.

Fred once attended a cooking-with-cheese class. He was worried that some day they might remove the vending machines from the hallway outside his office, and then he would need to learn to cook. He imagined that he would learn how to make cheese fondue or a cheese omelette. Instead the teacher stood up in front of the class with a list of cheeses.

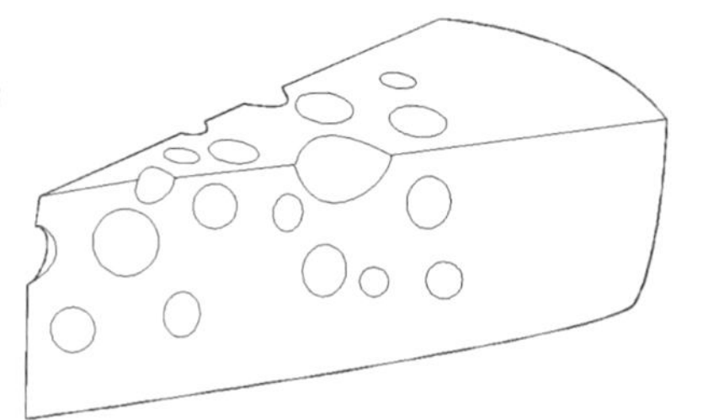

The teacher began reading: Acorn, Affidelice au Chablis, Ambert, American Cheese, Aragon, Armenian String, Aromes au Gene de Marc, Asadero, Baguette Laonnaise, Basket Cheese, Bavarian Bergkase, Beaufort, Bierkase, Bleu de Gex, Blue Castello, Boursault, Boursin, Bouyssou, Braudostur, Breakfast Cheese, Brebis du Lochois, Bresse Bleu, Brick, Brie, Brie de Melun, Briquette de Brebis, Briquette du Forez, Broccio, Bryndza, Buchette d'Anjou, Buffalo, Butte, Butterkase, Buxton Blue, Caboc, Caciotta, Caerphilly, Cairnsmore, Calenzana, Cambazola, Canestrato, Cantal, Caprice des Dieux, Capriole Banon, Carre de l'Est, Casciotta di Urbino, Cashel Blue, Castellano, Castelmagno, Castelo Branco, Castigliano, Cathelain, Cendre d'Olivet, Cerney, Chabichou, Chaource, Charolais, Chaumes, Cheddar, Cheshire, Chevres, Chontaleno, Civray, Coeur de Chevre, Colby, Comte, Coolea, Cooleney, Coquetdale, Corleggy, Cornish Pepper, Cotherstone, Cotija, Cougar Gold, Coulommiers, Coverdale, Cream Havarti, Crema Agria, Crema Mexicana, Creme Fraiche, Crescenza, Croghan, Crottin du Chavignol, Crowdie, Crowley, Cuajada, Cure Nantais, Curworthy, Cypress Grove Chevre. . . .

The hour was over by the time the teacher got to the end of the cheeses starting with *C*. "Your assignment," the teacher announced, "is to memorize the list. In the next class meeting, we will start with the cheeses beginning with *D*." Fred never attended the second class.

Also, he attended a world history from 1000 A.D. class. The teacher wrote on the board:

1009 – The first European paper mill is built.
1066 – The Normans conquer England.
1095 – First Crusade begins.
1206 – Everybody in Mongolia unites under Genghis Khan.
1248 – Roger Bacon describes the composition of gunpowder to Europe.
1258 – The Mongolians take over Baghdad.
1260 – The Mongolians take over China and Uzbekistan.
1275 – Marco Polo visits Kublia Khan in China.
1284 – Eye glasses invented.

Fred didn't wait around until 1300. He quietly left the classroom. He had learned some valuable lessons about teaching. He wrote them in his notebook so that he wouldn't forget them.

Good Teaching

1. *Making kids memorize facts is the easiest way to teach.*
2. *Making kids memorize facts is the worst way to teach.*
3. *Make it relevant. Who cares when Marco Polo went to China?*
4. *Make it fun.*
5. *A good teacher does not dispense facts like a vending machine dispenses soft drinks. After Gutenberg made books cheap back in the 1400s, straight lecturing of facts became stupid.*

Fred was always on the lookout for ways to play in the classroom. Learning should be a delight. When Fred spotted some other bookmarks in the store, he thought to himself How can I use these in the classroom? These bookmarks were made out of paper and only cost 5¢ each.

bookmark

He bought 82 of them—one for each person in his pre-algebra class. While Betty was doing some more shopping, Fred sat in the shopping cart and worked on his bookmarks.

On the first bookmark he wrote →

4 → 8
7 → 14
8 → 16
3 → 6

Fred was creating Function Bookmarks.™ A function is any fixed rule. This first bookmark had the rule *double the number.* When he gave the bookmarks to students, they were supposed to figure out what the rule was.

If he had written it would not be a function because 4 would have two different "answers."

not a function
4 → 8
7 → 14
4 → 16
3 → 6

A function is a fixed unchanging rule. The number 4 can't have an answer of 8 one day and have an answer of 16 another day.

Fred's second Function Bookmark looked like this:

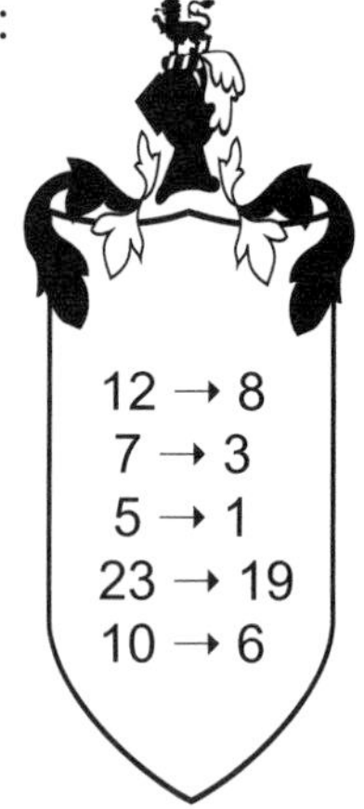

Look at it for a second and see if you can figure out what rule Fred was using.

Language Lesson

On this bookmark, Fred wrote 12 → 8.

12 → 8 is read "12 is **mapped** to 8."

12 → 8 is also read, "8 is the **image** of 12."

12 → 8 can be written as an **ordered pair** (12, 8).

This second bookmark that Fred wrote could have been written as a set of ordered pairs = {(12, 8), (7, 3), (5, 1), (23, 19), (10, 6)}. Fred was using the rule *subtract 4.*

You would never have a function that had these two ordered pairs in it: (55, 7) and (55, 39). Why? Because a function is a fixed rule that has only one "answer" (one image). If you had 55 → 7 and 55 → 39, then 55 would have two images.

Functions don't have to have numbers in them. Here is the third bookmark that Fred made:

As a set of ordered pairs, it would be

= {(Fred, black), (Joe, green), (Betty, pink), (Darlene, black)}

The rule was *name the one color you dislike the most.* This is a function. It is okay that black is the image of both Fred and Darlene. It would not be okay if we had both Joe → green and Joe → purple. Darlene would yell at him, "Joe! Make up your mind! Name the one color you dislike the most!"

Each thing must have exactly one image. That's a third definition of a **function**.

Your Turn to Play

1. You are given the function {(Harvard, no), (KITTENS, yes), (Stanford, no), (University of California at Berkeley, no), (Yale, no)}. What is the image of Stanford?

2. Continuing the previous problem, can you guess what the rule for this function is?

3. California → Sacramento, Idaho → Boise, Delaware → Dover is a function given by the rule *name the capital.* What is the inverse of this function? Inverses were talked about on pages 23, 38, and 117.

4. Give the inverse of the function {(3, 10), (45, 52), (0, 7)}.
(This is the function *add 7.*)

5. Explain why the function given in problem 1 does *not* have an inverse function.

6. $9\frac{2}{7} \div 2\frac{3}{5} = ?$

7. Guess the rule for this function: {(Fred, 1), (Betty, 2), (Alexander, 4), (Joe, 1), (Darlene, 2), (Taft, 1), (Washington, 3), (presbyopia, 5), (perambulation, 5)}. This is not easy.

Complete solutions are on page 200

Chapter Thirty
Graphing

Betty put Fred in a shopping cart while she picked out the books she wanted at Books! Books! Books! Bookstore. While Fred had been busy writing on his bookmarks, Betty had been piling books in the cart. When Fred disappeared under the pile of books, she thought it was time to check out at the cash registers.

The clerk took each of the books out of the shopping cart and scanned it. The clerk was half asleep. She had two full-time jobs, had three kids at home, and was an active member of the bowling league. Not paying much attention, she picked up Fred by his square head and tried to scan him. After several tries, she looked at what she had in her hand, noticed that he was a kid and not a book, and put him back into the cart. She looked at Betty and said, "Funny looking kid."

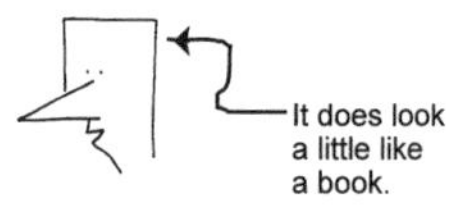

Betty had the books shipped to her apartment by KITTENS mail whose motto is "Faster than email." The books would arrive before Betty got home that night.

She and Fred walked out into the night air. She looked at Fred and said, "I'm sorry she called you 'funny looking.' That wasn't very nice."

He shrugged his little shoulders and said, "It's okay."

They walked for a while, and Fred explained his Function Bookmarks to Betty. She thought for a moment, took out a pen, and wrote on a slip of paper: (1, –1), (2, 8), (3, 19), (4, 31), (5, 43), (6, 52), (7, 56), (8, 55), (9, 44), (10, 32), (11, 18), (12, 5).

She handed it to Fred and said, "Some rules are a lot harder to guess. I bet that you can't guess my rule for this function."

Fred always loved a math challenge. The first thing that he did was check to make sure that it was a function. If Betty had listed two ordered pairs like (7, 56) and (7, 63), then it wouldn't have been a function.

It was a function.

Fred talked to himself, "One is mapped to minus one,* and two is mapped to eight."

"You know," he said to Betty, "sometimes it's easier to see if you turn it into a picture—if you graph the points. Take the ordered pair (2, 8), for example."

Another Language Lesson

(2, 8) is called an ordered pair.

2 is the **first coordinate**.

8 is the **second coordinate**.

2 is also called the **x-coordinate**.

8 is also called the **y-coordinate**.

Fred wanted to plot (2, 8) on a graph. He went over 2 and up 8 and made a mark.

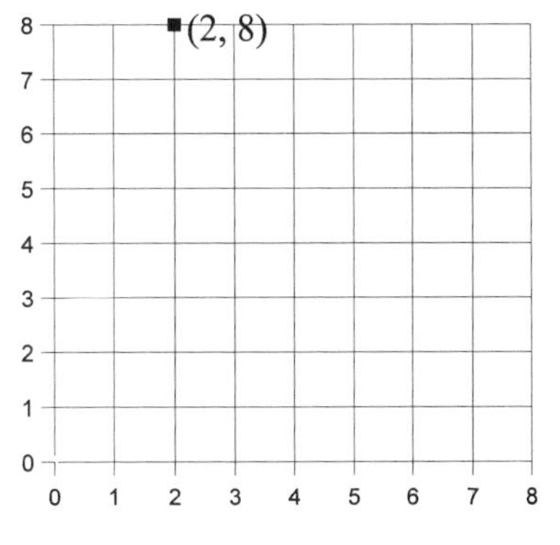

The first coordinate tells you how far to the right to go. The second coordinate tells you how far up to go.

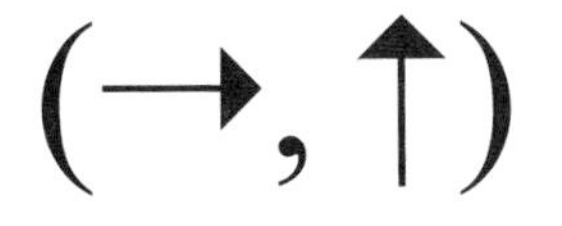

---

* Minus one (= −1) is a **negative number**. We have talked about the whole numbers, which are {0, 1, 2, 3, 4, . . .}. If we count downwards—4, 3, 2, 1, 0—why stop at zero? If we don't stop at zero, we get the integers, which are {. . . −3, −2, −1, 0, 1, 2, 3, 4, . . .}.

The negative numbers aren't very good for counting sheep. How can you have negative three sheep? But, of course, fractions aren't very useful for counting sheep either. How can you have $3\frac{1}{7}$ sheep?

But there are cases in which negative numbers make sense. If I have \$100, and I use my credit card to buy a \$99 textbook, then my net worth is \$1.

If I have \$100, and I use my credit card to buy a \$101 textbook, then my net worth is −\$1. I am worse than broke. If I find a dollar on the street, then my net worth climbs up to \$0.

Then Fred graphed all twelve points that Betty had given him.

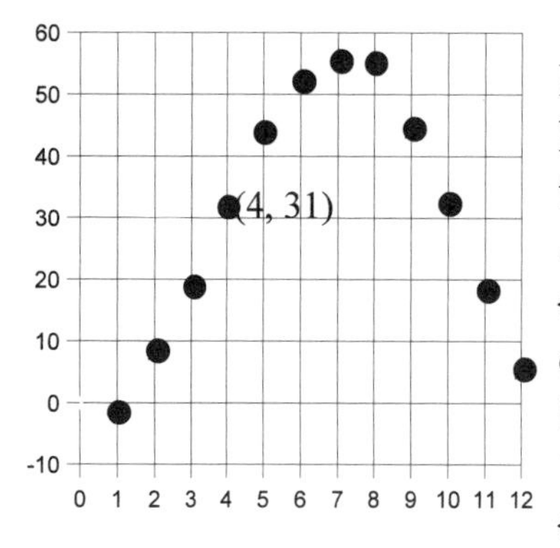

He stared at the graph. He scratched his head. Something is happening here he thought. There's a hump in the middle. And the first coordinates are from one to twelve. A dozen eggs? No. The twelve disciples? No. The twelve tribes of Israel? No. The twelve months of the year? Maybe. The twelve days of Christmas? Maybe. The twelve-step program of A.A.? Probably not.

"Do you give up?" Betty asked him. "The rule for this function is really hard to guess."

"Give me a hint," Fred said.

"Okay. The first coordinates are the months of the year."

Fred went into deep-think. Could it be the net profit at a department store? No, because you would expect the profit to be high during the Christmas season. Could it be the number of United States presidents born during each month? No, because you have the point (7, 56) and I'm sure 56 presidents weren't born in July. And besides, the point (1, -1) wouldn't make any sense.

They walked a couple more blocks, and Fred asked for the answer.

Betty said, "It's the average low temperatures for each month in Bismarck, North Dakota."

"I don't think I would have been able to guess that," Fred admitted.

Fred thought of something he would present in class tomorrow. "Can I try this out on you Betty?" he asked.

"Sure."

Fred drew three graphs and asked, "Can you tell which one of these three is *not* the graph of a function?"

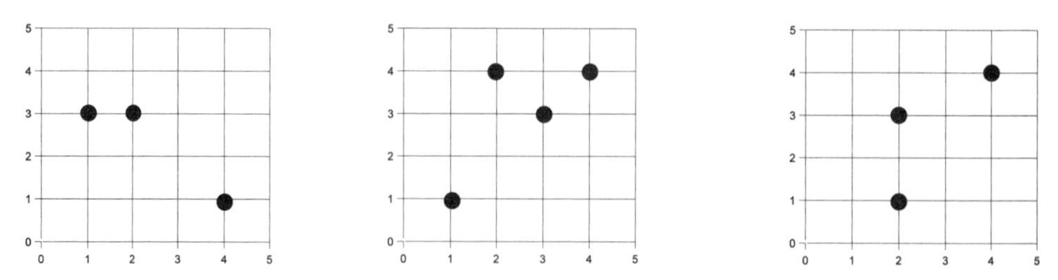

"That's easy," said Betty. "The third graph isn't a function."

Your Turn to Play

1. How did Betty see that the third graph wasn't a function?

2. One definition of function is: *A function is a set of ordered pairs in which no two different ordered pairs have the same first coordinate.*

   How can you quickly spot when a graph is not a function?

3. You may have noticed on the graph at the top of the previous page, the horizontal axis (the **x-axis**) was numbered 1, 2, 3 . . . , but the vertical axis (the **y-axis**) was numbered 10, 20, 30. . . . Draw a rough picture of what that function would look like if both axes* were numbered alike.

4. Which is colder: –1° or –8°?

Complete solutions are on page 201

✶ The plural of axis (AK-sis) is axes (AK-seize). English is weird.

The Bridge

from Chapters 1 – 30 to Chapter 31

first try

Goal: Get 9 or more right and you cross the bridge.

1. Fred can read Christina Rossetti's poems at the rate of 5.7 poems per hour. How many of her poems can he read in 1.8 hours?

2. Change $\frac{1}{5}$ into a percent.

3. Fred wanted to read 180 poems during the year. In January, he read 27 of them. What percent of the poems did he read?

4. Of the 27 poems he read in January, one-ninth of them he really liked.

How many did he really like?

5. A tear came in Fred's eye when he read:

> ***If we should meet one day,***
> ***If both should not forget,***
> ***We shall clasp hands the accustomed way,***
> ***As when we met,***
> ***So long ago, as I remember yet.****

He averaged two tears for every eleven poems he read. What is the ratio of tears to poems?

6. After he read each poem, he gave it a grade. "Goblin Market" got a *B*. "Maggie a Lady" got an *A*. "A Better Resurrection" got an *A*. Each of the 180 poems got a grade. This is a function. Does it have an inverse function?

7. {(Darlene, Joe)} is a function. What is the inverse function?

8. Find the area of the rectangle not covered by the two circles. (Use 3.14 for π.)

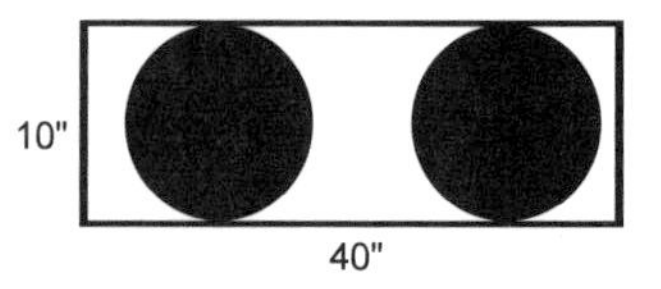

9. How many subsets does {☎, ⌚} have?

10. Change 10.02 into a mixed number. Remember to reduce fractions as much as possible.

✶ The third verse of Rossetti's "Twilight Night."

The Bridge

from Chapters 1 – 30 to Chapter 31

second try

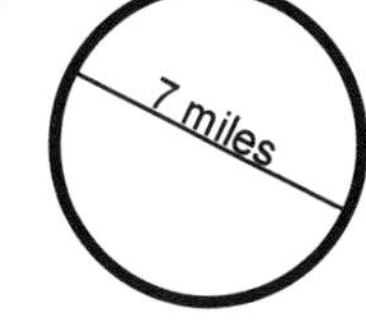

1. Joe thought he might have better luck fishing in a lake than in a swimming pool. He found a circular lake that was 7 miles across. What was its area? (Use 3.14 for π.)
2. Round your answer to the previous problem to the nearest tenth of a square mile.
3. Joe caught 3 fish. There are 3000 fish in the lake. What percent of the fish in the lake had he caught?
4. Four percent of the 3000 fish in the lake are guppies. How many guppies are in the lake?

5. Darlene was sitting in the boat with Joe. She didn't have much to do because Joe only brought one fishing pole. He suggested that she could pass the time by cleaning all the fish that he caught. One of the fish was a guppy. She cut off the head and tail and tossed them into the water. That decreased the weight of the 1.8-ounce fish by 35%. What was the weight of the fish with the head and tail gone?

6. Darlene classified each of the three fish that Joe had caught. Guppy → inedible. Lionfish → inedible. Platy → inedible. Is this a function?

platy

7. Is this a graph of a function?

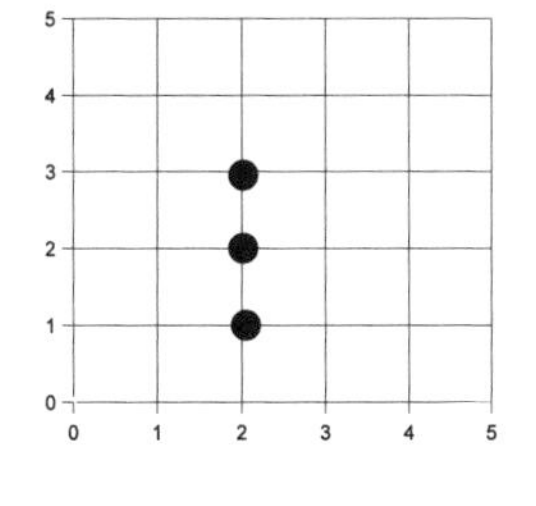

8. Is this true: $-3 < -5$?
9. Express 29 as the sum of three primes.
10. Is 11,000,000,000,003 evenly divisible by 3?

The Bridge

from Chapters 1 – 30 to Chapter 31

third try

1. In the cotton print section of the fabric store, Darlene found the perfect cloth to make Joe a giant handkerchief. (Joe always borrowed her handkerchief when his nose started to run.) She thought he would like the picture of the cowboy poet. She also liked the price. It was 25¢.

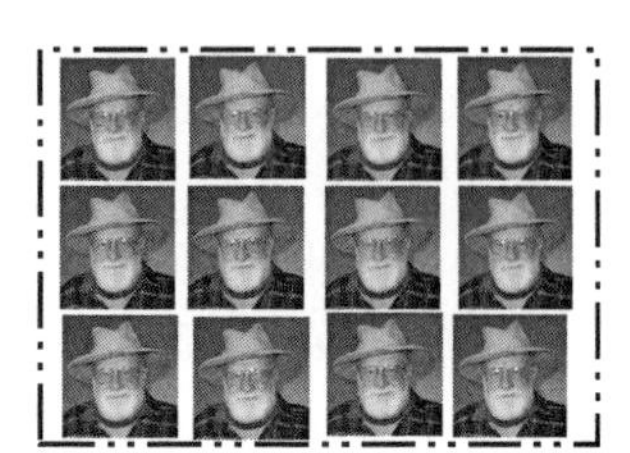

When she got home, she put it into the washing machine. The pictures washed off and the cloth lost its rectangular shape. What is the area of this parallelogram?

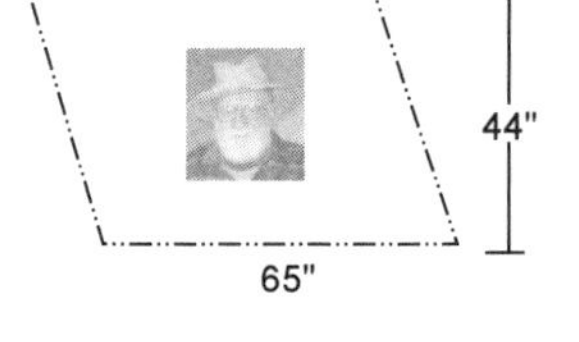

2. Change $\frac{3}{8}$ into a percent.

3. After Darlene had worked on Joe's handkerchief for 8 minutes, she estimated that she had completed 20% of the whole project. How long is the whole project?

4. Of the 30 sewing projects that Darlene started since she got her new sewing machine five years ago, she has completed 9 of them. What percent has she completed?

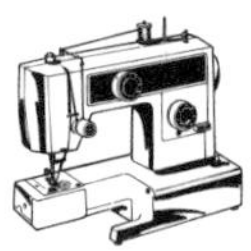

5. Continuing the previous problem, what is the ratio of completed sewing projects to uncompleted ones?

6. Many people who do a lot of sewing have a large collection of buttons. Darlene has only three buttons: her black button, her orange button, and her red button. She sorted the buttons: black → four holes, orange → two holes, red → four holes. Is this a function?

7. Continuing the previous problem, does it have an inverse function?

8. How many subsets does {❀, ✈, ✉} have?

9. Does $\frac{1}{6}$ terminate or repeat when it is written as a decimal?

10. Does the function *round to the nearest tenth* have an inverse?

The Bridge

from Chapters 1 – 30 to Chapter 31

fourth try

1. The sailor's wife had the bath mat cut down so that it would work as a place mat. She was careful not to cut off any part of the torch, but she didn't quite get things square. What is the area of the place mat?

2. Change $\frac{3}{4}$ into a percent.

3. When the five sailors were visiting New York, three of them decided to buy cameras and take pictures. What percent of the five bought cameras?

4. Of the 150 pictures that the three sailors took, 40% of them were duds. How many were duds?

5. Two of the five sailors got sick during their visit. What is the ratio of sick sailors to sailors that weren't sick?

6. {(first sailor, Statue of Liberty), (second sailor, Central Park), (third sailor, the camera store), (fourth sailor, Statue of Liberty), (fifth sailor, the Empire State building)} is a function that maps each sailor to his favorite place in New York. Does it have an inverse function?

7. Which of these are true? $-22 < -4$ $\quad 8 < 44$ $\quad -0.5 < 0.5$ $\quad -7 < 5$

8. Express 22 as the sum of three primes.

9. Change $\frac{7}{8}$ into a decimal.

10. The two sailors who got sick had both gone into the All-You-Can-Eat Ice Cream Store. The first sailor ate 0.79 kilograms of ice cream. The second sailor ate 7% more than the first. How much did the second sailor have?

The Bridge

from Chapters 1 – 30 to Chapter 31

fifth try

1. Kansas isn't noted for its earthquakes. Or for its beaches. But it is world famous for its KITTENS University. (And if it weren't for Fred Gauss, probably no one would have ever heard of KITTENS.)

Fred was writing an equation containing the first three letters of the Greek alphabet when he noticed that the blackboard wasn't quite square. (α = alpha, β = beta, γ = gamma.)

$\alpha^2 = \beta^2 - \gamma + 1$

37"

89"

What is the area of his blackboard?

2. Fred noticed that 15% of the 20 light bulbs in the classroom were burned out. How many were burned out?

3. A biology class meets in the classroom the hour before Fred's class. There were a lot of leftover frogs on the floor when he came into the room.

Of the 60 frogs, 40% of them were dead. What percent were not dead?

4. Continuing the previous problem, how many dead frogs were there?

5. What is the ratio of dead frogs to live frogs?

6. The biology teacher once told Fred that she buys frogs by the pound. A bucket of frogs weighs 39.01 pounds. If frogs weigh 0.47 pounds each, how many frogs are in a bucket?

7. Is this a function? {(Fred, blackboard), (frog, floor), (student, chair)}

8. Is this the graph of a function?

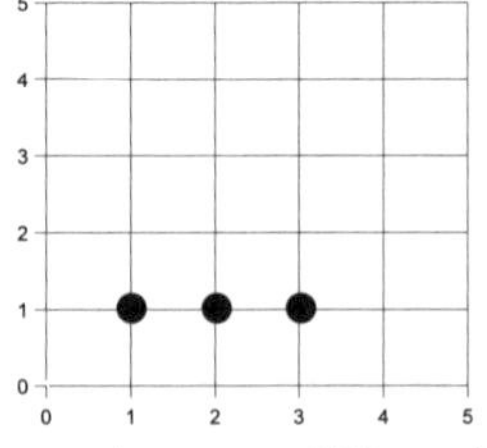

9. Write 69¢ using the "$" sign.

10. What is the smallest number larger than a million that is evenly divisible by 3 and 5?

They came to the edge of the KITTENS campus. Betty said goodnight to Fred and headed to her apartment. Fred headed toward his office. Walking on campus at night was a delight for Fred. He passed people engaged in quiet conversation. He could see the stars. The world smelled fresh—not like the 5100th Avenue Steak House.

Some people, when they walk in the evening, will recite lines from their favorite poetry.

> ***The night has a thousand eyes,***
> ***and the day but one;***
> ***Yet the light of the bright world dies***
> ***with the dying sun.***
>
> ***The mind has a thousand eyes,***
> ***And the heart but one;***
> ***Yet the light of a whole life dies***
> ***When love is done.****

Some recite the alphabet in Morse code: A •–, B –••• and some in Greek: α (alpha), β (beta), ...

The Nine Conversions popped into Fred's mind. These are the nine basic fraction–percent conversions that come in handy in everyday life. In the arithmetic class that Fred taught, he drilled the Nine Conversions every day for a couple of weeks until each student knew them as well as the multiplication tables.

He and his students would say together, "One-half equals fifty percent. One-third equals thirty-three and a third percent. . . ."

✶ "The Night Has a Thousand Eyes" by Francis William Bourdillion

Here are the Nine Conversions:

$\frac{1}{2}$ = 50%

$\frac{1}{3}$ = 33⅓%

$\frac{2}{3}$ = 66⅔%

$\frac{1}{4}$ = 25%

$\frac{3}{4}$ = 75%

$\frac{1}{8}$ = 12½%

$\frac{3}{8}$ = 37½%

$\frac{5}{8}$ = 62½%

$\frac{7}{8}$ = 87½%

These nine are probably the most useful ones to learn by heart. I also carry around in my head $\frac{1}{6}$ = 16⅔% and $\frac{5}{6}$ = 83⅓%, but if I added them to the above list, I couldn't call them the Nine Conversions anymore.

Almost everyone knows that $\frac{1}{10}$ = 10%, $\frac{2}{10}$ = 20%, $\frac{3}{10}$ = 30%, etc., so there was no reason to put them on the to-be-memorized list.

I don't know about you, but for me, memorizing lists is not my favorite way to spend an afternoon. We automatically memorize the things that we use a lot.

It isn't painful for someone interested in baseball to memorize the batting average of his favorite player.

It didn't take work for you to memorize where the kitchen and bathroom are where you live.

I know that π is approximately 3.14159, because I use it frequently in my life.

While you are memorizing $\frac{1}{8}$ equals 12½%, I want to talk about your future a little bit. In a couple of weeks (or less), you will graduate from *Life of Fred: Decimals and Percents.* You will have finished arithmetic. This is an MTP (major turning point) in your life.

Some MTPs in your life are ones that you can hardly remember—for example, when you were born. Or the day that you discovered you could play with your toes. Or the time you discovered that it was fun to play with other kids. (And you noticed that playing with other kids was different than playing with your toes. It was a definite improvement.*) Finishing arithmetic is an MTP for several reasons. First, after finishing the book you are now holding in your hands, you will know more mathematics than most people who have ever lived.

It was only about 400 years ago (in 1624) that Henry Briggs invented the modern method of long division:

$$\begin{array}{r} 807 \\ 683\overline{)\,551181} \\ -\ \underline{5464} \\ 4781 \\ -\ \underline{4781} \end{array}$$

And the average person on the street today can't answer the question: 60 is what percent of 480?

60 = ?% of 480

60 ÷ 480 (The number closest to the "of" is the divisor.)

$\frac{60}{480}$

which reduces to $\frac{1}{8}$

which is 12½%.

This could also be solved by $480\overline{)\,60.000}$ with quotient 0.125 = 12.5%

Another reason why finishing arithmetic is a major turning point is that much of the plain, old memorizing is behind you. There is a lot of mathematics ahead of you, but 90% of it you won't have to memorize.

---

* . . . at least, for most people.

When I taught math at the college level, the arithmetic course was the only course in which there were closed-book tests. In the tests in all the courses that followed arithmetic, I allowed the free use of books, notes, calculators*— but, of course, not cell phones!

Here is your future. After arithmetic (fractions, decimals, and percents) are: ❀ Pre-Algebra

❀ Beginning Algebra

❀ Advanced Algebra

❀ Geometry

❀ Trigonometry, and then

❀ Calculus.

Finish these subjects and you will be at the junior (the third-year) level at a university in mathematics. At that point, most universities will offer you the chance to check a box: ☒ **I wish to declare mathematics as my major.**

Wait! Stop! I, your reader, have a silly question.

✶ This is not a universal practice at the college level. Many teachers in math courses after arithmetic still have closed-book exams. Remember the first rule of good teaching (from two chapters ago): *1. Making kids memorize facts is the easiest way to teach.*

I wanted my students to spend their time learning *how* to do the math and not spend that time learning how to be tape recorders.

Out there in the "real world" (as it's called), you need to know that $7 \times 8 = 56$. It would be embarrassing or inconvenient to haul out your calculator to get that answer.

But, if you are an engineer needing to find the indefinite integral $\int \frac{dx}{16 - x^2}$ your boss isn't going to say to you, "Now close your book and do the problem." Instead, you will head to your office, close the door, pour a strawberry milkshake, and pull a Table of Integrals off your shelf. It will tell you that the answer is $\frac{1}{8} \log \frac{4 + x}{4 - x} + C$. (The "indefinite integral" is from calculus. The "log" is from advanced algebra.)

I believe that after arithmetic, what is important to know will be there, not because you have sat down and memorized it, but because you have used it often enough that it will memorize itself. When you were learning your first language, did you work with a vocabulary list and memorize the meaning of words? Of course not.

I bet I know what you are going to ask. You want to know how I can write a *Your Turn to Play* to find out if you have memorized these Nine Conversions.

Yup. How did you know what I was thinking? You seem to be able to read my mind. Okay, Mr. Author, show me how you are going to do this. You can't just write, "Problem 1: Convert $\frac{5}{8}$ to a percent." If you did, I could just look at the table above and have the answer. Good luck writing the *Your Turn to Play* **!**

Your Turn to Play

1. I . . . um . . . $\frac{5}{8}$ . . . phooey

. . . . . . .COMPLETE SOLUTIONS . . . .

1.

You are right, my reader. I can't think of a way to test whether or not you have memorized that $\frac{7}{8}$ equals 87½%. It's like phoning someone up and asking them to shut their eyes and tell me what color shirt they are wearing.

I guess we will have to do it on the Honor System. Without looking back, please do this *Your Turn to Play*.

Your Turn to Play

1. Convert each of these to a percent: $\frac{1}{3}$, $\frac{5}{8}$, $\frac{1}{4}$, $\frac{1}{2}$, $\frac{1}{8}$
2. Convert each of these to a percent: $\frac{2}{3}$, $\frac{7}{8}$, $\frac{3}{8}$, $\frac{3}{4}$

Complete solutions are on page 201

Fred climbed the stairs to the third floor of the Math Building. He walked past the vending machines in the hallway. *I've had enough to eat today. I'm not really hungry.* He entered his office and turned on the light. He stepped on something. It was a little gear. He picked it up and threw it into the wastebasket. On his desk was part of a list that he had made that morning. That also went into the wastebasket.

Why I Want a Bike
page 2
5. I am no longer a baby. I used to be 5, but now I'm 5½.
5 < 5½
6. I can walk at 3 mph. On a bike I can go 10 mph. Everyone knows that 3 < 10. Riding is faster.

His day had begun at 6 A.M. After saying his morning prayers, he got up, rolled up his three-foot sleeping bag, dressed, and headed down the hallway to see what was for breakfast. He purchased a candy bar and a cup of Sluice from the vending machines and took them back to his office. He never got around to consuming them.

Now it was 10 P.M. He had been awake for . . .

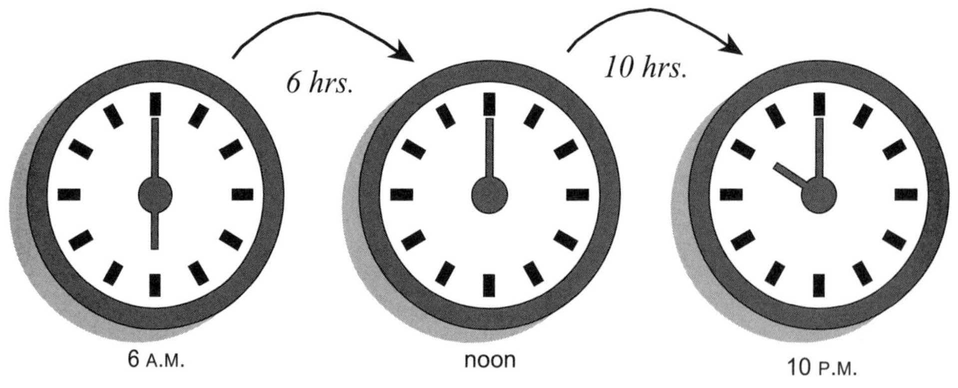

. . . he had been awake for 16 hours. That is a long time for a five-and-a-half-year-old boy. Sixteen hours out of 24 is $\frac{16}{24}$ which reduces to $\frac{2}{3}$ which is $66\frac{2}{3}\%$ of the day.

Fred flossed his teeth to clean up after the one-eighth of an ounce of marshmallow that he had eaten earlier during the day. Fred was never considered a fast eater. He liked to chew his food. He had started "dining" on his marshmallow at 11:52 A.M. and had finished at 12:17 P.M. It had taken him . . .

. . . it had taken him 25 minutes to eat it.

Fred had read in some book* that you should floss for as long as it took you to eat. After every minute of flossing, the string would start to fray, and Fred would pull out a fresh foot of string. Tonight, he would need 25 feet of string.**

The "odometer" on his box of floss said that there was only 6 feet left.*** After he finished that box, he would use . . .

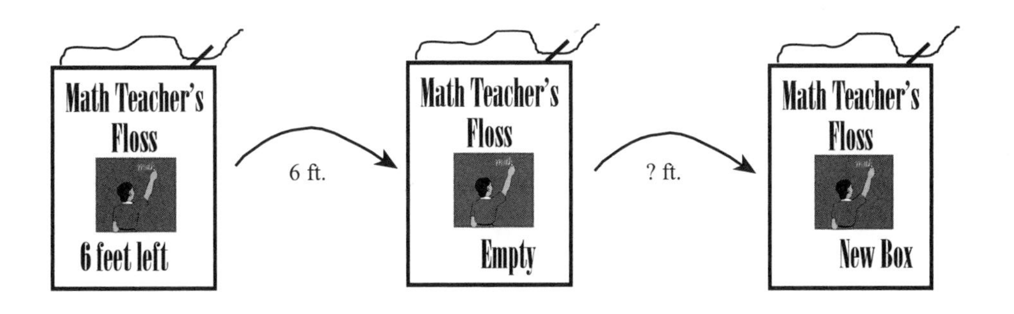

. . . 19 feet from the new box. (25 – 6 = 19)

---

* Prof. Eldwood's *Flossing for the Modern Man*, 1849. Believing everything you are told can sometimes be a mistake.

** One of Fred's worries was that some day he might have to attend an eight-course banquet. Then he might be flossing till dawn.

*** Besides his many awards for teaching excellence, Fred had been declared Floss Master by the American Tooth & Gum Society. With that award came a little gadget that attaches to a box of floss. It is called a Floss-o-dometer. It tells you how much floss is left in the box.

Your Turn to Play

1. How long is it between the time you arrive at the doctor's office (11:48 A.M.) and when you are seen by the doctor (12:26 P.M.)?

2. There is 0.08 of an ounce of toothpaste in the old tube. Fred uses that up and then uses some from a new tube. Because he had eaten $\frac{1}{8}$ of an ounce of marshmallow, he used $\frac{1}{8}$ of an ounce of toothpaste. How much did he use from the new tube?

3. $8\frac{1}{3} + 1\frac{4}{5} = ?$

4. In *Life of Fred: Advanced Algebra Expanded Edition,* Fred took a bus trip that started at 12:10 P.M. and ended at 5 P.M. two days later. How many hours did he travel?

5. Fred, Peggy, Alexander, and Betty were on a picnic at the Great Lake on the KITTENS campus. (This happened in *Life of Fred: Statistics Expanded Edition.*) They were all having fun naming functions. Peggy suggested *Point to the tallest person here at the picnic.* She pointed to Alexander. List this function as a set of ordered pairs. The domain and the codomain are both {Fred, Peggy, Alexander, Betty}.

Complete solutions are on page 201

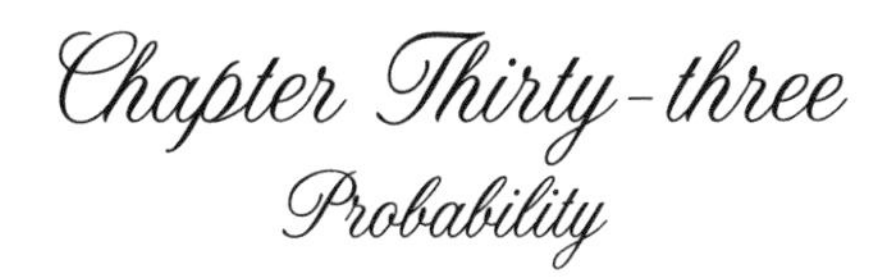

Chapter Thirty-three
Probability

With teeth as clean as his, Fred didn't need a nightlight. All he had to do was smile, and the room was filled with a soft light. This also came in very handy on the nights (99.9% of them) that he wanted to read in bed before falling asleep.

Fred unrolled his three-foot sleeping bag underneath his desk. He selected about a dozen books from his bookshelves and placed them beside his sleeping bag. He changed into his pajamas, turned off the light, and smiled his way to his sleeping bag. There in bed he read by the light of his smile.

This worked well for happy sections of books, but when he got to places like the last chapter of *A Man Called Peter* in which Peter Marshall dies and steps over into the Larger Life, Fred began to cry, and the light went out. On those occasions, he used a flashlight.

Fred's habit was to reach over without looking and pick a book out of his stack. That added a little excitement to his bedtime rituals.

Tonight he had twelve books in his stack, and three of them were history books. What was the probability that when he selected a book, it would be a history book? There were 3 chances in 12. That's $\frac{3}{12}$ or $\frac{1}{4}$ or 25%.

The probability that it would *not* be a history book was $\frac{9}{12}$ or $\frac{3}{4}$ or 75%.

In tonight's stack of twelve books, he did not have Hilda Doolittle's *Selected Poems*. What was the probability that when he selected a book from the stack, it would have been Hilda Doolittle's *Selected Poems*? There were 0 chances in 12. That's $\frac{0}{12}$ or 0%.

In contrast, every one of the books in tonight's stack was longer than 150 pages. What is the probability that a book selected at random out of that stack would be more than 150 pages? There was 12 chances in 12. That's $\frac{12}{12}$ or 100%.

Eight of the books in tonight's stack were originally written in English.✶ What is the probability that when Fred selected a book at random from the stack it would be one that was originally written in English? There are 8 chances in 12. That's $\frac{8}{12}$ or $\frac{2}{3}$ or 66⅔%.

✶ One indication that your reading is "growing up" is the number of books you have read that weren't originally written in English. The world is large, and great things have been written in many different languages. If none of the books you have read has ". . . translated by . . ." on the title page, you are missing a lot. It is like being an 18-month-old who just plays in the living room, while all the older kids are playing outside. When I was in high school, our class read *The Odyssey* (Greek).

In the past five hundred years, starting with Dante's *The Divine Comedy* (Italian), there has been great literature written in several languages. For example, Pascal's *Pensées* (French), Goethe's *Faust* (German), Cervantes's *Don Quixote* (Spanish), and Erasmus's *In Praise of Folly* (Latin).

And if you look at the last hundred years or so, important writers are found everywhere. For example, José Saramago's *Baltasar and Blimunda* (Portugal), Joan Perucho's *Natural History* (Catalonia), Boris Pasternak's *Doctor Zhivago* (Russia), Sigrid Undset's *Kristin Lavansdatter* (Scandinavia), and Michael Thelwell's *The Harder They Come* (the West Indies). The list goes on and on.

And, of course, the book that has had more editions printed than any other book, was written in Hebrew and Greek. (The Bible.)

Where to start? In most libraries there are books that list the great books. My favorite is the third edition of Clifton Fadiman's *Lifetime Reading Plan.* Because *Lifetime Reading Plan* was written in 1988, it naturally does not include any of the books in the *Life of Fred* series.

Your Turn to Play

1. Fred has 24 students in his set theory class. Three of them are freshmen. If he picks a student at random from that class, what is the probability that the student will be a freshman? Give your answer both as a fraction and as a percent.

2. Two of the three freshmen are female. If you select a freshman at random, what is the probability that the student will be female? Give your answer both as a fraction and as a percent.

3. What is the smallest probability possible? Give an example of something with that probability.

4. Could an event have a probability of 120%?

5. $8\frac{1}{3} \times 1\frac{4}{5} = ?$

6. How much time is there between 8:12 A.M. and 3:41 P.M.?

Complete solutions are on page 202

Fred started to get sleepy. It had been a long day, a day that began way back in *Life of Fred: Fractions* when he made his list: "Why I Want a Bike." He had taught in the classroom filled with television cameramen, photographers, and a sculptor. He had headed off to C.C. Coalback's bicycle store and had purchased a bike that cost every cent he had in his checking account. When the bicycle box arrived, he tried to open it with a huge knife and managed, instead, to injure himself.

Betty took him to the hospital where he was fixed up. Alexander joined them, and they all headed off to have lunch at PieOne. Fred got a job as a pizza cook, met a little lamb, got fired, and went to get cleaned up at a car wash. He headed back to his office to open his bicycle box.

Fred got a pet to play with, but Billy Bug didn't last long. A much larger (1") playmate was Bobbie Bot. When Fred, Betty, and Alexander headed off to the 5100th Avenue Steak House for dinner, Bobbie constructed Berta and Elizabot. After a tea party, the three little robots built a not-so-small* brother robot named Roger. While he was sitting and playing in the middle of the KITTENS football field, the United States military forces turned Roger into dust.

Fred went walking in the twilight hours. He spent a couple of moments in the C.C. Coalback Toy Store but couldn't find anything to buy. He spent more time in the Books! Books! Books! Bookstore, but too much of that time was spent trapped in the children's section of the store. Betty showed up at the bookstore and aided Fred's escape.

Fred headed back to his office and got ready for bed. After reading for a while, he set his books aside. He thanked God for the full day he had had, for his friends, and for his teaching position at KITTENS. He turned over and was soon fast asleep.

You, dear reader, are almost done with arithmetic. You have the final bridge to cross. As usual, you have five tries. I wish you the best of luck.

---

* An example of litotes (LIE-teh-tease). Litotes is almost the opposite of hyperbole. It is litotes to say that Roger was "not-so-small" when he was taller than the Chrysler Building.

Litotes is the negative of the contrary. For example, if you are wearing a bathing suit in the middle of a blizzard in Montana, you might say, "I am not feeling very warm right now."

The Final Bridge

All 33 chapters

first try

Goal: Get 17 or more right and you cross the bridge.

1. Which base system do we use: the base five system, the base ten system, or the base twenty system?
2. Does 55.0700 equal 55.07?
3. If the radius of a circle is equal to 2.2 yards, what is its circumference? (Use 3.14 for π.)
4. Is this a function? *Assign ★ to the number of pages in today's newspaper.*
5. Suppose that A and C are sets. What does $A \subset C$ mean?
6. The books in Fred's office weigh 839.8 pounds. The university president decided that all of the office walls should be painted sunshine yellow. Painting the walls in Fred's office was a real waste of money because he had so many books that you couldn't see them.

The movers took his 839.8 pounds of books and carried them off in 17 loads. All the loads weighed the same. What was the weight of a load?

7. When the books were removed from Fred's shelves, the painters discovered that the walls had never been painted. The walls were the original white plaster. (The Math Building had been constructed in 1929.) Someone had scribbled on the wall Of the fractions $\frac{1}{2}, \frac{1}{3}, \frac{1}{4}, \frac{1}{5}, \frac{1}{6}, \frac{1}{7}, \frac{1}{8}, \frac{1}{9}$, and $\frac{1}{10}$ only $\frac{1}{3}, \frac{1}{6}$, and $\frac{1}{9}$ are repeating decimals. Is this true? If so, write "yes." Otherwise, name the other fraction or fractions that repeat when expressed as a decimal.
8. Another scribble was π^2 = ?
If you use $\pi = 3.1$, what is the value of π^2?
9. List all the composite numbers between 61 and 71.

Final Bridge—*First Try*—continued on next page.

The Final Bridge

All 33 chapters

10. One wall of Fred's office was 17.9 feet wide and 8 feet tall. What was its area?

11. After the painter finished painting that wall, he drew a giant happy face on the wall. He made it as large as he could. What was the area of that happy face? (Use 3.14 for π.)

12. The painters estimated that it would take 8 gallons to paint Fred's office. They used 13% of that on the happy-face wall. How much paint did they use on the happy-face wall?

13. If the painters were painting at the rate of 1.3 gallons per hour, how much paint would they use in 1.3 hours?

14. It normally takes 1.6 hours to paint a wall that is 8 feet by 17.9 feet, but because of the scribbles on the wall, it takes 27% longer. How many hours will it take to paint that wall?

15. What is the perimeter of this triangle?

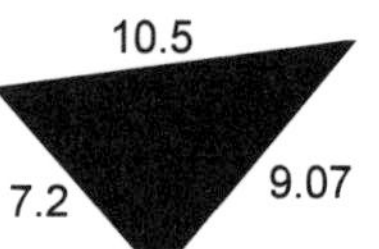

16. By trial and error, find $\sqrt{5041}$.

17. One of the painters was only four years old, and could only paint the bottom 5 feet of the 8-foot-tall wall. What percent of the wall could he paint?

18. Continuing the previous problem, what is the ratio of painted to unpainted wall?

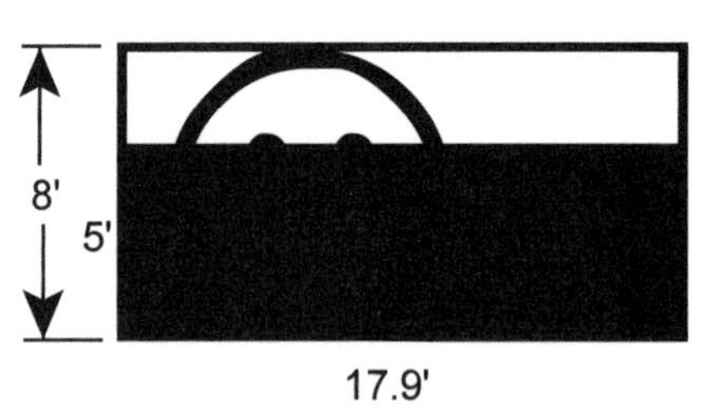

19. Graph these ordered pairs {(2, 3), (1, 4), (3, 3)}.

20. Continuing the previous problem, is {(2, 3), (1, 4), (3, 3)} a function?

The Final Bridge

All 33 chapters

second try

---

1. List the whole numbers.
2. 39.9 – 39.09 = ?
3. Round 829.555 to the nearest hundreth.
4. Darlene and Joe decided to take his boat out into the ocean to do some deep-sea fishing. Actually, Joe decided. Darlene thought that the rowboat was too small to take out into the ocean. During the first hour that they were out, it was raining at the rate of 1.1" per hour. During the second hour the rate increased by 20%. How fast was it raining during the second hour?

5. Suppose that set B = {7, ❀, green} and set D = {6, 7, ❀}.
   Then B – D = ?
6. In four hours, the outgoing tide carried Joe and Darlene and their little rowboat 32.36 miles out to sea. How fast was the tide flowing? Assume that the tide was flowing at a constant rate.
7. Express $\frac{2}{3}$ as a decimal, rounding your answer to the nearest thousandth.
8. Express $0.\overline{6}$ as a percent. (The answer is not 60%.)
9. List all the prime numbers between 70 and 80.
10. If a quart of ocean water cost 0.25¢, how many quarts could you buy for a dollar?
11. Thirty-five percent of all people who are 32.36 miles out to sea in a rowboat end up in a hospital. If there were 700 people in rowboats 32.36 miles out to sea, how many of them would end up in a hospital?
12. The waves grew higher and higher. Joe had 3,700 fishing lures with him. He liked to tell Darlene, "You can never have too many lures." When the boat rocked, he lost 6% of them overboard. How many lures did he have left?
13. Joe had 12 fishing rods. He lost one of them overboard. What percent of his fishing rods did he lose?

Final Bridge—*Second Try*—continued on next page.

14. The waves were 0.7 feet high before the storm came up. When the storm came up, they were 89% higher. How high were the waves during the storm?

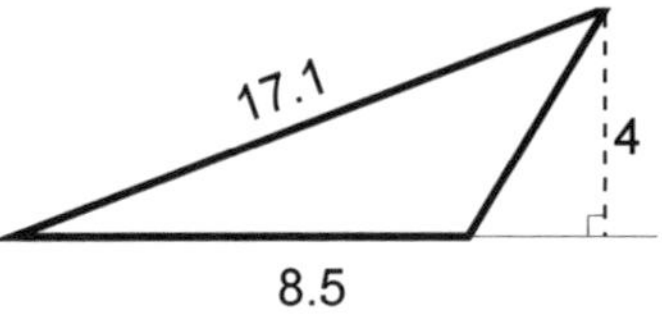

15. What is the area of this triangle?

16. The rowboat began to fill with water. The rowboat was 32" tall, and the bottom 12" was filled with water. What percent of the boat was filled?

17. Is {(7, 77), (8, 77), (9, 77)} a function?

18. Graph {(7, 77), (8, 77), (9, 77)}.

19. Just as the boat was beginning to fill with water, Joe caught his first fish. He pulled it onboard and asked Darlene to take the hook out of its mouth. She dropped the fish into the bottom of the boat, and it began to swim around in the water around their ankles. At that point, Darlene started to worry about whether they would survive. The fish that Joe caught looked happier than Darlene did. She said, "Joe, dear, I think it's time we started heading back to shore. Do you mind if I row while you fish?"

Joe said, "Sure,*" and Darlene began to row. She started at 11:45 A.M. and rowed until 3:15 P.M. At that point her arms and back were too tired to row anymore. How long had she rowed?

20. By 3:15 P.M. Joe had caught seven more fish, and the water in the boat had reached their waists. Of the eight fish now swimming around inside their boat, five of them were platy fish. If you were to select a fish at random from the eight fish swimming around in Joe's boat, what is the probability that it would be a platy? Express your answer as a percent.

---

* Joe was answering the question, "Is it okay if I row?" The answer to that question is "Yes" or "Sure." Because Darlene asked, "Do you mind if I row?" the answer should be "No" or "I don't mind if you row."

The Final Bridge

All 33 chapters

third try

1. Three score is equal to how much in the base ten system?

2. What is a thousand times 0.08?

3. Darlene wanted to make a flag for Joe's boat. She cut out a piece of cloth in the shape of a parallelogram. She chose the shape of a parallelogram rather than the traditional rectangle because she thought it would look like the flag was waving. What is the area of that flag?

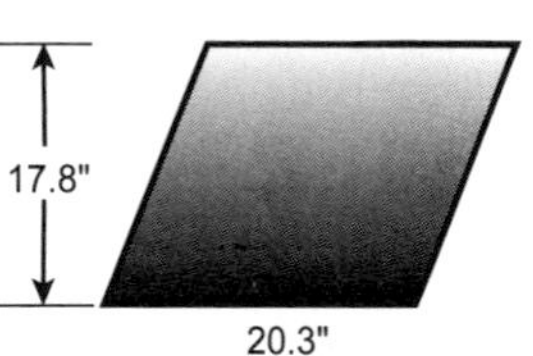

4. Darlene had six yards of red thread on a spool. She used two feet. How much did she have left on the spool?

5. Suppose you have a number that is *not* evenly divisible by 3. If you multiply that number by 1,000,000, will the answer be evenly divisible by three? Choose one of these answers:

A) Yes. It will always be evenly divisible by 3.

B) Sometimes it will be, and sometimes it won't be. It depends on the number with which you started.

C) No. If the original number was not divisible by 3, then multiplying by 1,000,000 will never give you a number evenly divisible by 3.

6. Joe didn't like a flag in the shape of a parallelogram. He wanted a circular flag with a circumference of 21 inches. Darlene asked, "How are you going to put that on a flag pole?" Joe shrugged. He also indicated that he wanted a picture of a fish on the flag. What is the diameter of the flag? Use $\pi = 3.14$ and round your answer to the nearest tenth of an inch.

Joe's drawing of the flag he wanted

7. List the composite numbers that are less than 17.

Joe's new drawing

8. Joe changed his mind. He wanted a circular flag with a picture of a fisherman on it. He told Darlene that the radius of this new flag should be 8 inches. What would be its area? Use 3.14 for π.

Final Bridge—*Third Try*—continued on next page.

The Final Bridge

All 33 chapters

9. Change $\frac{1}{9}$ into a decimal and round your answer to the nearest thousandth.

10. Thirty percent of the time, Joe wanted a flag with a fish on it. Sixty percent of the time, he wanted a picture of a fisherman on his flag. Ten percent of the time, he wanted a plain blue flag. Draw a pie chart to illustrate this situation.

11. Joe has made 156 fishing trips in the last six months. Twenty-five percent of them have been disastrous. How many of them have been disastrous?

12. Darlene went to the fabric store to buy the materials to make a flag for Joe. The satin cloth was normally \$4.60 per yard. The sign said, "Sale! 30% off!" What was the sale price?

13. She used 0.17 yards of thread each minute. How many yards would she use in an hour?

14. It takes Darlene 3.7 hours to sew a rectangular flag, but it takes 44% longer to sew one that is circular. How long does it take her to sew a circular flag?

15. Darlene didn't want to make a circular flag. It was too much trouble. She suggested to Joe that a pennant would be nice. He didn't know what a pennant was, so Darlene took a pencil and made a small pennant to show him. Joe liked the idea and drew a diagram of the pennant he wanted.

a small pennant

What was the area of Joe's pennant?

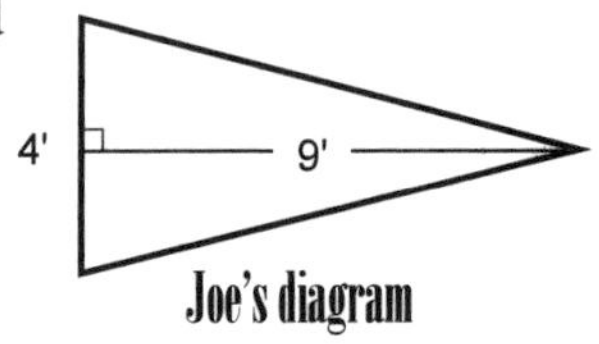

Joe's diagram

16. By trial and error find $\sqrt{4096}$.

17. Darlene had 48 square yards of facing material, and she used 42 square yards of it in making Joe's flag. What percent did she use?

18. Continuing the previous problem, what fraction of her 48 square yards did she use? Reduce your answer.

19. Darlene used 42 square yards of fabric to make flags for Joe because Joe kept changing his mind about the flag he wanted. She had used the 42 square yards to make 16 flags. Joe liked two of the flags, and Darlene threw the rest into the garbage can. What percent of the flags that she made ended up in the garbage?

20. What is the ratio of flags Joe liked to those he didn't like?

The Final Bridge

All 33 chapters

fourth try

---

1. 9.09 + 33. + 0.07 = ?

2. What is the inverse function of *subtract sixteen*?

3. Statue of Liberty dolls were selling for $6.78 each. The sailors bought nine of them. How much was the total cost?

4. One sailor was 6'5" tall. Another was 5'7" tall. How much taller was the first than the second?

5. If the sum of the digits of a number add to 12, is the number evenly divisible by 5? Choose one of these answers:

A) Yes. It will always be evenly divisible by 5.

B) Sometimes it will be, and sometimes it won't be. It depends on the number with which you started.

C) No. It will never be evenly divisible by 5.

6. Change 0.207 into a fraction.

7. The first sailor (Jake) had $70 in his pocket. The second sailor (Jack) had $10, and the third sailor (Jane) had $50. Draw a bar graph to illustrate how much money Jake, Jack, and Jane had in their pockets.

8. Find three prime numbers that add to 31.

9. Jane had a coin from the country of Freedonia in her pocket. It had a picture of a famous math teacher on its face. What was the area of that coin? (Use 3 for π.)

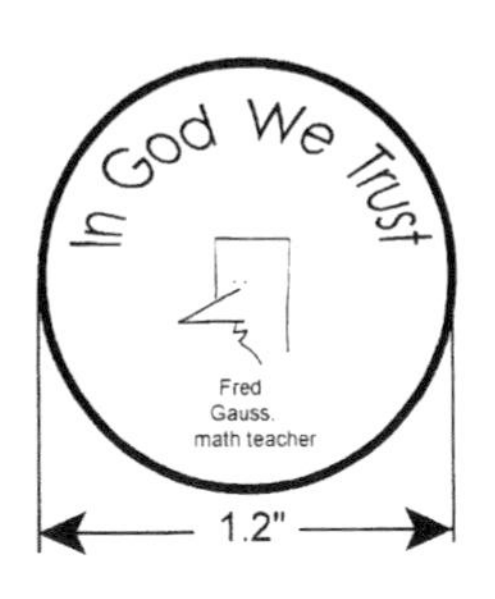

10. Jake said that Jane's Freedonia coin was worth 0.50¢. If that was true, how many of those Freedonia coins could you buy for $3.00?

11. Jake estimated that there were 1,300 possible places to eat lunch in New York. Jane said that 19% of those places served hamburgers, and that she was in a mood for a hamburger. How many places could they choose from if they were going to satisfy Jane's desire for a hamburger?

12. Jake, Jack, and Jane went to the 6300th Avenue Hamburger Shop. The menu included the 6300th Avenue Hamburger Shop's famous 1.3-pound

Final Bridge—*Fourth Try*—continued on next page.

hamburger. There was an asterisk attached to the 1.3-pound. The footnote read: "1.3 pounds before cooking." Jack asked the waiter how much was lost in cooking. "30%" was the answer he received. What would be the net weight of the hamburger after cooking?

13. Jake liked to estimate things. He estimated that he was spending about $9.80 each hour he was in New York with his friends. How much would he spend in 5.7 hours?

14. Jake spends $9.80 per hour, and Jack spends 7% more per hour. How much does Jack spend per hour? Round your answer to the nearest cent.

15. Jake ordered a hamburger lasagna. He failed to read the fine print on the menu that said This lasagna is served in a rectangular dish and will easily feed six hungry people. When it arrived, he gasped. He took a knife and cut it along a diagonal and put half of the lasagna in his duffle bag. Here is a diagram of the lasagna that remained on his plate. What is the perimeter?

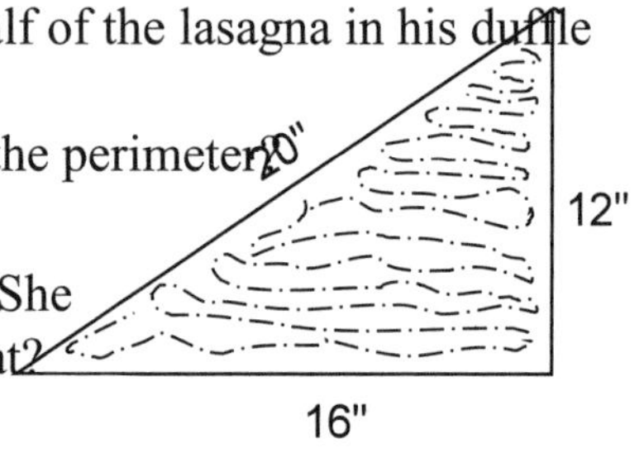

20"
12"
16"

16. What is the area of that lasagna?

17. Jane's hamburger weighed 0.9 pounds. She ate 0.6 pounds of it. What percent did she eat?

18. What is the ratio of the part she ate to the part she didn't eat?

19. The sailors had left their ship at 7:50 A.M. and had to be back by 11:45 P.M. How much time did they have to see New York?

20. When they got back on ship, they decided to have a midnight snack before heading off to bed. Jake took the 6.09 pounds of lasagna that he had saved from lunch out of his duffle bag . They divided it equally among the three sailors. How much did each receive?

The Final Bridge

All 33 chapters

fifth try

---

1. 4.05 – 2.007 = ?

2. If the diameter of a circle is equal to 7 inches, what is its circumference? Use 3.14 for π.

3. If Fred used chalk at the rate of 0.08" per minute, how much would he use in an hour?

4. The blackboard was almost 4 yards wide. In fact, it was 3 inches less than 4 yards wide. How wide was it? Express your answer in inches.

5. Is 555,550,000,000,777 evenly divisible by 3?

6. In his 55-minute presentation of the History of Mathematics, Fred used 49.5 ounces of chalk. How much chalk did he use each minute? Assume that his rate of chalk use was constant.

7. List all the prime numbers between 60 and 70.

8. Find two prime numbers that add to 32.

9. Fred drew a large square on the blackboard. It was 24.2 inches on each side. What is the area of the largest circle that could be drawn inside that square? (Use 3 for π.)

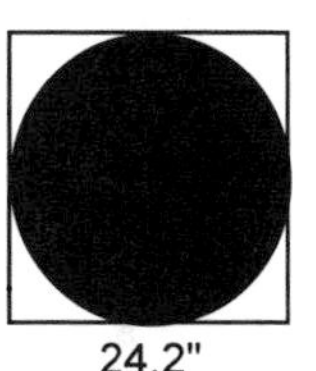

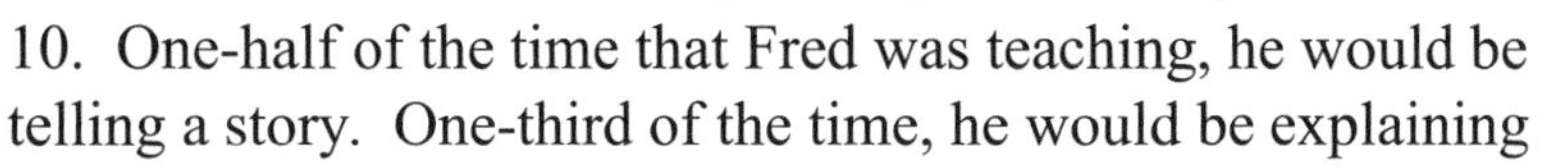

10. One-half of the time that Fred was teaching, he would be telling a story. One-third of the time, he would be explaining the math. One-sixth of the time, he would be answering questions that the students asked. Draw a circle graph to illustrate Fred's use of time in the classroom.

11. Other teachers at KITTENS would often come and sit in Fred's classroom. They enjoyed hearing the stories Fred would tell. One-seventh of the 161 people in Fred's geometry class were teachers. How many teachers were in his geometry class?

12. Fred took an English class from Mrs. Angemessen. On one essay that Fred wrote for that class, Mrs. Angemessen wrote, "I was going to give you 97.3 points out of a possible 157 points, but I took 40% off because you were too funny." How many points did Fred receive on that essay?

Final Bridge—*Fifth Try*—continued on next page.

The Final Bridge

All 33 chapters

13. Fred had to write a term paper for Mrs. Angemessen's class. He was assigned the topic. "What I Did Today." She gave him strict instructions not to be "too funny." He knew this was going to be a hard paper to write. He wrote at the rate of 0.81 pages per hour. How many pages could he write in 4.6 hours?

14. In his essay "What I Did Today," Fred spent 40 pages describing his purchase of a bicycle. The section describing his employment at PieOne was 10% longer. How many pages was his description of his employment at PieOne?

15. Mrs. Angemessen told her students to take out a piece of paper and write a paragraph on "The Importance of Ice Hockey in the Development of Panamanian Culture." Fred couldn't think of anything to write. He stared at the blank page. What was the area of this 8.5" × 11" sheet of paper?

16. By trial and error find $\sqrt{1849}$.

17. Fred looked at the clock. Mrs. Angemessen had given the students 48 minutes to write the paragraph. Twelve minutes went by, and Fred's paper still had only his name on it. What percent of the 48 minutes had elapsed?

18. Mrs. Angemessen required that the paragraph be 250 words long. Fred wrote, "I really don't know what the importance of ice hockey has been in the development of Panamanian culture." That was 18 words. What percent of the required paragraph had Fred written?

19. Here is a five-digit number with one of the digits missing: 7?825. What must the missing digit be in order for the number to be evenly divisible by 9?

20. Round 7,395,039 to the nearest million.

The Bridge
answers

from p. 29 — *first try*

1. The vigesimal system is base 20.
2. 109.86
3. 453.025 grams
4. 0.0000072 grams
5. $44° \times \dfrac{60 \text{ minutes}}{1°} = 2640$ minutes
6. The inverse function of *divide by a million* is *multiply by a million.*
7. $C = \pi d$, so if $d = 10$, then $C = \pi 10$ or 10π.
8. 1340. (or 1340)
9. $0.09 \doteq 0.1$ gram
10. A thousand times heavier than 0.00072 grams is 0.72 grams.

from p. 30 — *second try*

1. 2.09 divided by a thousand is 0.00209
2. The inverse function of *adding six* is *subtracting six.*
3. 8.102
4. 62 = 3 twenties + 2 ones = **32** in the vigesimal system.
5. $8.956 \doteq 8.96$
6. $200.17 - 8.003 = 192.167$ grams
7. One-tenth of 10.07 is 1.007 grams.
8. $10.07 \doteq 10.1$ grams
9. Ten times 0.3¢ is 3.¢ (or 3¢).
10. A million times 10.07 moves the decimal 6 places to the right. 10,070,000. grams

from p. 31 — *third try*

1. 437.652
2. Dividing 88 by a million moves the decimal 6 places to the left. 0.000088
3. The inverse function of *losing 4 degrees of temperature* is *gaining 4 degrees of temperature.*
4. Ten times \$1.07 is \$10.70.
5. $98.6 - 0.07 = 98.53°$
6. One hundred times 992.3 is 99230. grams (or 99230 grams or 99.23 kg)
7. $120 \text{ minutes} \times \dfrac{1°}{60 \text{ minutes}} = 2°$
8. $992.3 \doteq 992$ grams (or 992. grams)
9. One millionth of 992.3 = 0.0009923 grams
10. $4.4399 \doteq 4.44$

from p. 32 — *fourth try*

1. The inverse function of *multiply by 9.4* is *divide by 9.4.*
2. 36.97
3. $856.7724 \doteq 860.$ (or 860)
4. To divide 93.2 by a billion, you move the decimal 9 places to the left. 0.0000000932
5. $\pi \doteq 3.14$
6. $360 \text{ seconds} \times \dfrac{1 \text{ minute}}{60 \text{ seconds}} = 6$ minutes (or 6')

The Bridge
answers

7. 72.05 ≐ 72.1 mph
8. 3.417
9. One hundredth of 98 is 0.98 lbs.
10. 98 – 1.2 = 96.8 lbs.

from p. 33 — *fifth try*
1. 16.313
2. To divide 0.439 by a million, you move the decimal 6 places to the left. 0.000000439
3. 873.00389 ≐ 873.0039
4. $5° \times \frac{60 \text{ minutes}}{1°} = 300$ minutes
5. 12.3" – 2.08" = 10.22" lost
6. 46.02
7. One-thousandth of 43. is 0.043 minutes.
8. One-hundredth of $29. is $0.29 (or 29¢)
9. 4. – 2.2 = 1.8 increase in diopters
10. 3,539,982.007 ≐ 4,000,000 (or 4,000,000.)

from p. 52 — *first try*
1. 3.14 × 7.4 = 23.236 inches
2. The inverse of a function undoes what the function did. There is no way to "uncut" a slice of Aunt Alice's muffin. Once you've cut it, it's cut. There is no inverse.
3. 2 cups – 3 oz. = 1 cup + 8 oz. – 3 oz. = 1 cup 5 oz. or 16 oz. – 3 oz. = 13 oz.
4. The Aunt Alice raisin muffin is not a whole number so we would use ∉.
5. None of them divides evenly into 780001.
6. 70.05 grams
7. 7 × 4.75 = 33.25 grams
8. 329.09
9. 300
10. 42.98

from p. 53 — *second try*
1. $A - B$ = {Jackie, Sig}. Those are the elements that are in A and not in B.
2. They all do.
3. 813.01
4. 0.056007
5. 6.4 kg
6. 3.14 × 14.2 = 44.588 inches
7. *Lower the pizza three inches.*
8. 3 lbs. 7 oz. or 55 oz.
9. Zero or 0.0
10. 8977.7688

The Bridge
answers

from p. 54 — *third try*

1. 0.1925
2. It would be really tough to take the gas out of your car and shove it back into the gas pump. The idea of an inverse is that it restores everything to its original state. This action doesn't have an inverse.
3. 44' 3" – 4' 9" = 43' 15" – 4' 9" = 39' 6"
4. W = {0, 1, 2, . . .} and the smallest member of that set is 0.
5. W doesn't have a largest element. If it did, you could just add one to that number and you'd have an even larger whole number.
6. 5.0802
7. 21.96 gallons of gas
8. 3.14 × 0.077 = 0.24178 inches
9. You move the decimal six places to the left. 47. becomes 0.000047
10. Only 5 divides evenly into it.

from p. 55 — *fourth try*

1. A diameter is twice the radius. 2 × 20.3 = 40.6 feet
2. 3.14 × 40.6 = 127.484 feet
3. As far as I know there is no way to undo *throwing a pancake into the water.* There is no way to restore it to its original state. If you claim that you could take the pancake out of the water, dry it off, and claim that it would be just as nice and edible as when Darlene first gave it to Joe, then you could say that the action has an inverse. But in my experience pancakes that are thrown into the water turn yucky and mushy and can't be restored. So I'll say that the action has no inverse.
4. 3 cups – 7 oz. = 2 cups + 8 oz. – 7 oz. = 2 cups 1 oz. or 24 oz. – 7 oz. = 17 oz.
5. $J - D$ is the set of all elements of J that are not in D. Since none of the things that Joe is interested in are things that Darlene is interested in, $J - D = J$.
6. 8.3 – 5.8 = 2.5 kg
7. 60.84
8. 3.0 (or 3)
9. Multiplying by a billion moves the decimal 9 places to the right. 4,500,000,000.
10. Only 3 divides into it evenly.

from p. 56 — *fifth try*

1. It has an inverse. It is *multiply by* π *and then subtract 13.5.* That will return you to your original number.
2. 3 miles – 67 feet = 2 miles + 5280 feet – 67 feet = 2 miles 5213 feet
3. 1.24 – 0.9 = 0.34 m
4. 645,000 ÷ 1,000,000.
5. 3.14 × 0.67 = 2.1038 feet
6. Only 2 divides into it evenly.
7. 7.3 × 15 = 109.5 jelly beans
8. The elements that are in B that are not in P are just the elements of B. $B - P = B$ = {Bobbie}
9. To divide by a million move the decimal 6 places to the left. 0.0000007
10. 50.92

The Bridge
answers

from p. 79 — *first try*

1. 17.8 lbs.
2. 48 minutes
3. $\frac{73}{100}$
4. $44\frac{73}{100}$
5. 0.057 pounds
6. $0.1\overline{6}$
7. $0.\overline{285714}$
8. He could set 206 tables.
9. Zero is the only whole number that is not a natural number.
10. 9 years, 51 weeks

from p. 80 — *second try*

1. 150.2 grams
2. 62.4 ounces
3. In order for a number to be divisible by 9, the sum of its digits must be divisible by 9.
The sum of eighteen *2*s is 36. That is divisible by 9.
So the original number is divisible by 9.
4. $7\frac{4}{5}$ oz.
5. 0.125
6. 2400 slices
7. 2000 slices
8. 7.54 lbs.
9. 3.9"
10. 0.2"

from p. 81 — *third try*

1. 47 sheep
2. $3257\frac{1}{10}$ yards
3. 36.8 lbs. or $36\frac{4}{5}$ lbs.
4. 0.2
5. 46.3 hours
6. 3300
7. $119.38
8. 803.
9. 56.14
10. 3.9347

from p. 82 — *fourth try*

1. $6\frac{39}{50}$
2. Since the last digit is zero, it is divisible evenly by five.
3. 49.3

4. \$82.41
5. \$16.48
6. \$17.59
7. 0.05
8. Twenty billion or 20,000,000,000.
9. 6 hours, 54 minutes, 40 seconds
10. 1 hour, 11.1 minutes

from p. 83 — *fifth try*

1. \$13,119.12
2. \$52.06
3. 15.13 minutes
4. $8\frac{9}{10}$ minutes
5. 534 seconds
6. 0.0045 miles
7. The digits that we know in 879?205345 add up to 43. Replace the ? by 2 and they add up to 45 which is divisible by 9.
8. 0.906
9. 49 yards, 2 feet, 11 inches
10. $0.\overline{571428}$

from p. 99 — *first try*

1. 30, 32, 33, 34, 35, 36, 38, 39, 40
2. 3 + 37 = 40 or 11 + 29 = 40 or 17 + 23 = 40
3. The radius of the circle is 7 miles. The area of the square is 196 square miles.
4. 12.5%
5. 1%
6. $6^2\pi - 4^2\pi = 36 \times 3 - 16 \times 3 = 108 - 48 = 60$ square inches
7. \$0.087777, which rounds to 9¢
8. $7\frac{3}{1000}$
9. 8
10. *Add 7 and then divide by 3.* In algebra, we will write this as $f^{-1}(x) = \frac{x+7}{3}$

$f^{-1}(x)$ is read "f inverse of x."

from p. 100 — *second try*

1. 23 + 29 = 52 or 11 + 41 = 52 or 5 + 47 = 52
2. 300 square feet
3. Assuming the signs are correct, the second store is selling its oranges for only $\frac{1}{2}$¢ per pound. That is much cheaper than the first store's 37¢ per pound.
4. $\frac{1}{10} = 0.1 = 10\%$
5. $\frac{1}{50}$
6. 705 shreds of coconut. (282¢ ÷ 0.4¢/shred = 705 shreds)

7. 13.45 oz.
8. 414 verses (The sum of the digits of 414 is divisible by 9.)
9. There are four possible answers: { }, {Joe}, {Darlene}, or {Joe, Darlene}.
10. 11.56

from p. 101 — *third try*

1. The sum of its digits will have to be divisible by 3, and the number will have to be even. 1,000,000,002.
2. 23 + 31 = 54 or 7 + 47 = 54 or 11 + 43 = 54 or 13 + 41 = 54 (but not 1 + 53, because 1 is not a prime number)
3. d = 32 feet. r = 16 feet. $A = \pi r^2 = 3.14 \times 16 \times 16 = 803.84$ square feet
4. $803.84 \doteq 804$ square feet
5. \$3.40
6. $\frac{4}{5}$
7. $\frac{1}{20} = 0.05 = 5\%$
8. \$64.74 ÷ 830 yards = \$0.078 per yard (or 7.8¢ per yard)
9. B has four subsets. Any set containing two elements will have four subsets.
10. The subsets of C are: { }, {x}, {y}, {z}, {x, y}, {x, z}, {y, z}, and {x, y, z}. There are eight subsets.

from p. 102 — *fourth try*

1. 40, 42, 44, 45, 46, 48, 49, and 50
2. 9.42 feet
3. 7.065 square feet
4. 7.1 square feet
5. 5000¢ (or 5,000¢ or 5000.¢)
6. 6.3 – 0.7 = 5.6 lbs.
7. $\frac{19}{20}$
8. (See diagram →)

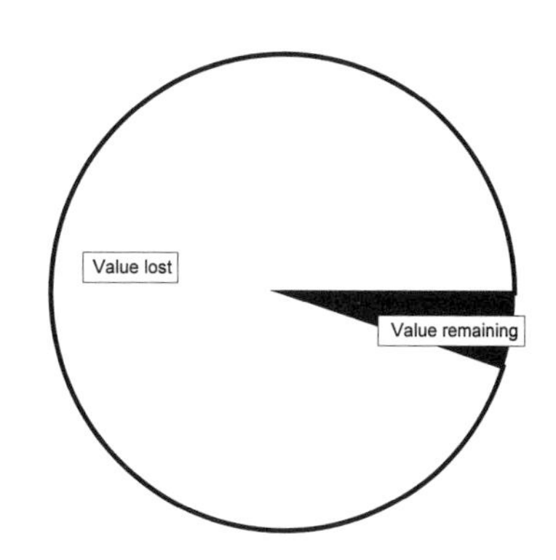

9. $\frac{3}{8} = 0.375 = 37.5\%$ (or $37\frac{1}{2}\%$)
10. 13 + 43 = 56 or 3 + 53 = 56 (but not 5 + 51 because 51 is not a prime number)

from p. 103 — *fifth try*

1. Each circle has an area of 254.34 square inches. The area of the rectangle is 900 square inches. 900 – 254.34 – 254.34 = 391.32 square inches
2. 80¢ (or 80.¢)
3. \$0.09 is largest
4. $\frac{7}{20} = 0.35 = 35\%$ (or 35.%)

The Bridge
answers

5. $\frac{17}{25}$
6. 3 × 477 = 1431 grams
7. 91%
8. 61and 67
9. 3 + 7 + 11 = 21 or 7 + 7 + 7 = 21 or 2 + 2 + 17 = 21
10. $43\,\frac{203}{1000}$

from p. 120 — *first try*

1. 3.36 hours
2. 34 books were nonfiction
3. 1323 books haven't been lent.
4. 9.9 yards
5. 3.5 × 6 = 21 poems
6. 1.2 × 6 = 7.2 poems per hour
7. 39
8. 23.1 square inches
9. 5.63
10. $7^\circ \times \frac{60 \text{ minutes}}{1^\circ} = 420$ minutes

from p. 121 — *second try*

1. $6.67
2. $13.02
3. 3%
4. 43.8 inches
5. 3.65 feet ≐ 4 feet
6. 1.3 × 0.67 = 0.871 fish/hour
7. (½)(30)(15) = 225 square feet
8. 29 + 37 = 66 or 7 + 59 or 13 + 53 or 19 + 47 or 5 + 61
9. $3 \times 4.5^2 = 60.75$ square inches
10. 81

from p. 122 — *third try*

1. 48.5 yards
2. 576 purple sequins
3. 58.96 ≐ 59 revolutions per minute
4. 19"/min × 1½ hrs. = 19"/min × 90 minutes = 1710 inches
5. 142.5 feet ≐ 143 feet
6. 2.7 × 1.71 = 4.617 sequins/minute
7. 179.55 square inches
8. 2 and 3
9. zero
10. 66

The Bridge
answers

from p. 123 — *fourth try*

1. 0.69 lbs.
2. 89%
3. $50
4. $4.71
5. $1.08 \times 1.57 = 1.6956 \doteq \$1.70$
6. 62.5%
7. $3 \times 1.5^2 = 6.75$ square feet
8. $3.14 \times 30^2 = 2826$ square miles
9. $7 + 23 = 30$ or $11 + 19 = 30$ or $13 + 17 = 30$
10. 57

from p. 124 — *fifth try*

1. 8 sticks/hr × 90 minutes = 8 sticks/hr × 1.5 hours = 12 sticks
2. 2901 students
3. $39
4. 50%
5. $1.3 \times 97.50 = \$126.75$
6. 34
7. 12.18
8. 50.24 square feet
9. 34.98
10. 182 square feet

from p. 144 — *first try*

1. 10.26 poems
2. 20%
3. 15%
4. 3 poems
5. 2:11 or $\frac{2}{11}$
6. No it doesn't. If it did, then A would be assigned to both "Maggie a Lady" and "A Better Resurrection."
7. {(Joe, Darlene)}
8. 243 square inches
9. There are four subsets of {☎, ⌚}. They are { }, {☎}, {⌚}, and {☎, ⌚}.
10. $10\frac{1}{50}$

from p. 145 — *second try*

1. 38.465 square miles
2. 38.5 square miles
3. 0.1% $\quad 3000\overline{)3.000}$ with quotient $.001$ $= 0.1\%$
4. 120 guppies
5. 65% of 1.8 = 1.17 ounces
6. Yes. Each fish has exactly one image.

The Bridge
answers

7. No, because 2 has more than one image. It is mapped to 1, 2, and 3.
8. No.
9. 5 + 11 + 13 = 29 or 5 + 5 + 19 or 5 + 7 + 17 or 3 + 13 + 13 or 7 + 11 + 11
10. No. The sum of the digits equals 5.

from p. 146 — *third try*

1. 2860 square inches
2. 37½% or 37.5%
3. (8 = 20% of ?) 40 minutes $0.20\overline{)8.00}$ $\begin{array}{r}40\\20\overline{)800}\end{array}$
4. (9 = ?% of 30) 30%
5. 9:21 or $\frac{9}{21}$ or $\frac{3}{7}$
6. Yes. Each button is assigned exactly one image.
7. No. The inverse would have four holes assigned to both black and red.
8. 8 subsets. They are: { }, {❀}, {✈}, {✉}, {❀, ✈}, {❀, ✉}, {✈, ✉}, and {❀, ✈, ✉}.
9. $\frac{1}{6}$ = 0.16666 . . . = $0.1\overline{6}$ is a repeating decimal.
10. No. The function *round to the nearest tenth* would map both 35.12 and 35.13 to 35.1. The inverse would map 35.1 to more than one image.

from p. 147 — *fourth try*

1. 837 square inches
2. 75%
3. 60%
4. 60
5. 2:3 or $\frac{2}{3}$
6. No. The Statue of Liberty would have two images: the first and fourth sailors.
7. All of them are true.
8. 2 + 7 + 13 = 22 or 2 + 3 + 17 = 22
9. 0.875 $\begin{array}{r}.875\\8\overline{)7.000}\end{array}$
10. 1.07 × 0.79 = 0.8453 kgs.

from p. 148 — *fifth try*

1. 3293 square inches
2. 3
3. 60%
4. 24
5. 40:60 or 4:6 or $\frac{40}{60}$ or 24:36 or $\frac{24}{36}$ or $\frac{2}{3}$ or 2:3
6. 83 $0.47\overline{)39.01}$ $\begin{array}{r}83\\47\overline{)3901}\end{array}$
7. Yes.
8. Yes.
9. \$0.69 or \$.69
10. 1,000,005

The Final Bridge
answers

from p. 161 — *first try*

1. We use the base ten system.
2. Yes.
3. 13.816 yards
4. No it isn't. In order to be a function, each time you apply the function you should get the same answer.
5. A ⊂ C means that A is a subset of C. Or you could say that A ⊂ C means that every element of A is an element of C.
6. 839.8 ÷ 17 = 49.4 pounds
7. $\frac{1}{7}$ is also a repeating decimal.
8. 3.1 × 3.1 = 9.61
9. 62, 63, 64, 65, 66, 68, 69, 70.
10. 143.2 square feet
11. 50.24 square feet
12. 0.13 × 8 = 1.04 gallons of paint were used on the happy-face wall
13. $\frac{1.3 \text{ gallons}}{1 \text{ hour}} \times \frac{1.3 \text{ hours}}{1} = 1.69$ gallons of paint
14. 1.6 × 1.27 = 2.032 hours
15. 26.77
16. 71
17. 62.5% or 62½%
18. 5 to 3 or 5:3 or $\frac{5}{3}$ or $\frac{(5)(17.9)}{(3)(17.9)}$
19. (See the diagram →)
20. Yes, it is a function.

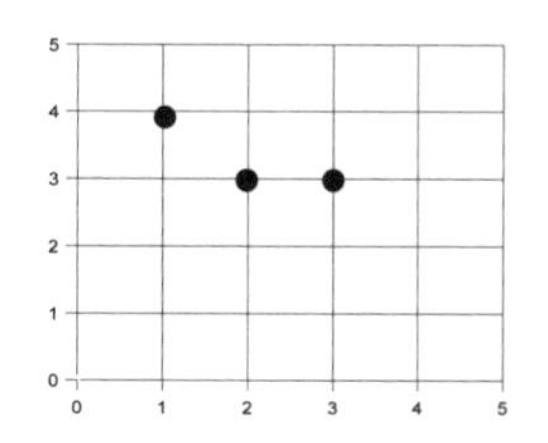

from p. 163 — *second try*

1. {0, 1, 2, 3, . . . }
2. 0.81
3. 829.56
4. 1.1 × 1.20 = 1.32" per hour
5. B – D = {green}
6. 32.36 ÷ 4 = 8.09 miles per hour
7. $\frac{2}{3}$ = 0.6666666. . . which rounds to 0.667
8. $0.\overline{6}$ means 0.6666666. . . which is equal to $\frac{2}{3}$ by the previous problem, which is equal to 66⅔% by the Nine Conversions of Chapter 31.
9. 71, 73, 79.
10. 100¢ ÷ 0.25¢ = 400 quarts
11. 700 × 0.35 = 245 people in the hospital.
12. 0.94 × 3,700 = 3,478 lures.
13. $\frac{1}{12}$ = 8⅓% or $8.\overline{3}$% or 8.333...%
14. 0.7 × 1.89 = 1.323 feet
15. 17

The Final Bridge
answers

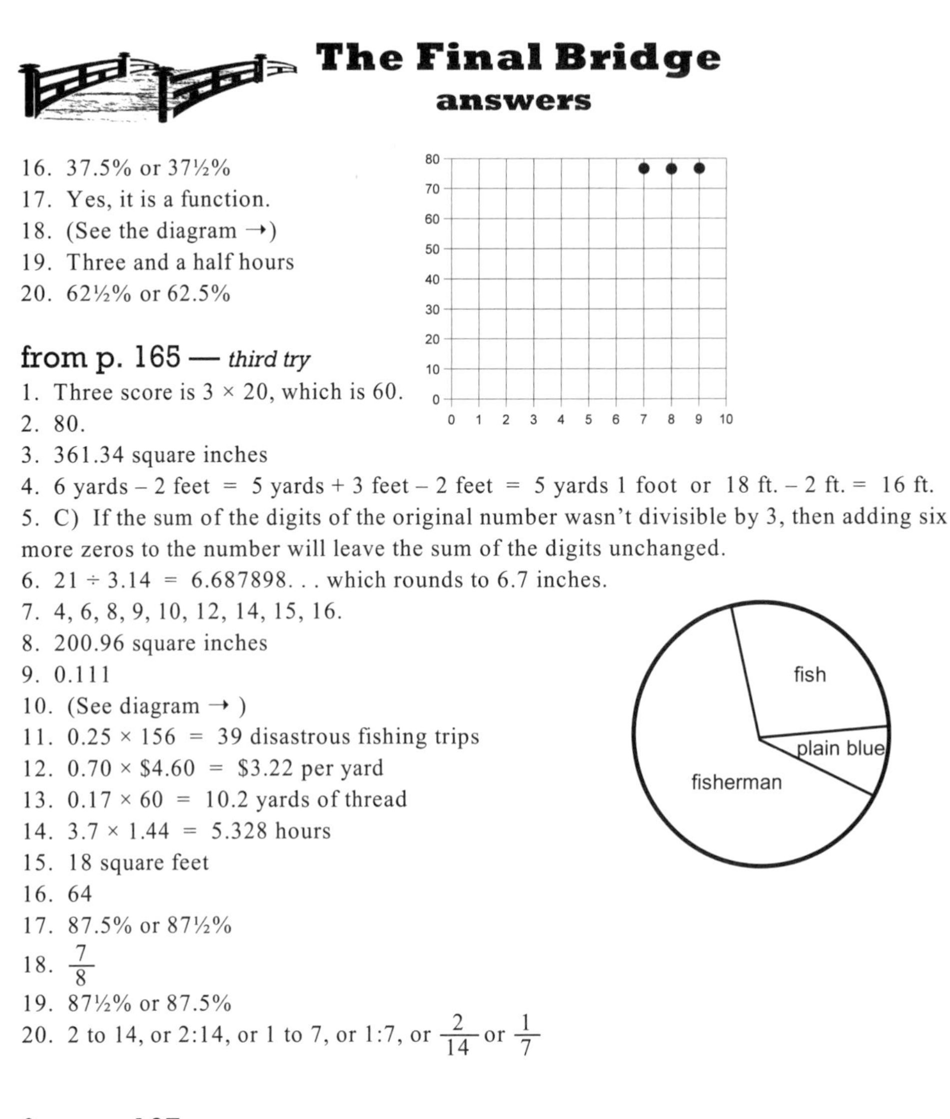

16. 37.5% or 37½%
17. Yes, it is a function.
18. (See the diagram →)
19. Three and a half hours
20. 62½% or 62.5%

from p. 165 — *third try*

1. Three score is 3 × 20, which is 60.
2. 80.
3. 361.34 square inches
4. 6 yards – 2 feet = 5 yards + 3 feet – 2 feet = 5 yards 1 foot or 18 ft. – 2 ft. = 16 ft.
5. C) If the sum of the digits of the original number wasn't divisible by 3, then adding six more zeros to the number will leave the sum of the digits unchanged.
6. 21 ÷ 3.14 = 6.687898. . . which rounds to 6.7 inches.
7. 4, 6, 8, 9, 10, 12, 14, 15, 16.
8. 200.96 square inches
9. 0.111
10. (See diagram →)
11. 0.25 × 156 = 39 disastrous fishing trips
12. 0.70 × $4.60 = $3.22 per yard
13. 0.17 × 60 = 10.2 yards of thread
14. 3.7 × 1.44 = 5.328 hours
15. 18 square feet
16. 64
17. 87.5% or 87½%
18. $\frac{7}{8}$
19. 87½% or 87.5%
20. 2 to 14, or 2:14, or 1 to 7, or 1:7, or $\frac{2}{14}$ or $\frac{1}{7}$

from p. 167 — *fourth try*

1. 42.16
2. The inverse function of *subtract sixteen* is *add sixteen.*
3. $61.02
4. 6'5" – 5'7" = 5' 17" – 5'7" = 10"
5. B) It depends on the number you started with. 48 isn't evenly divisible by 5, but 480 is.
6. $\frac{207}{1000}$
7. (See diagram →)
8. 3, 11, 17 or 7, 7, 17 or 5, 13, 13 or 7, 11, 13
9. 1.08 square inches
10. 300¢ ÷ 0.50 = 600 Freedonia coins
11. 0.19 × 1300 = 247 lunch places in New York that serve hamburgers.

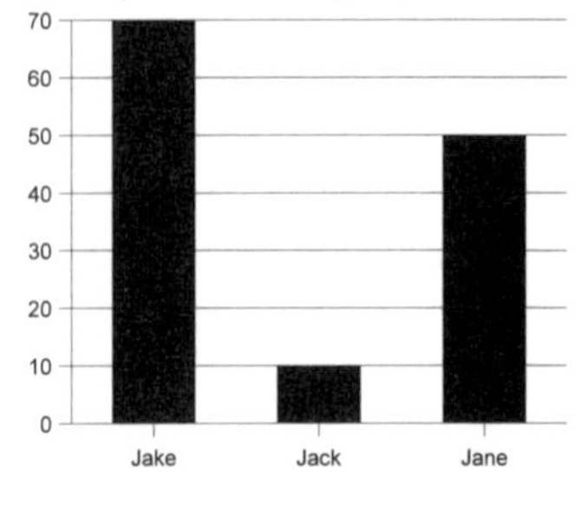

The Final Bridge
answers

12. $0.70 \times 1.3 = 0.91$ pounds
13. $\dfrac{\$9.80}{1 \text{ hour}} \times \dfrac{5.7 \text{ hours}}{1} = \$55.86$
14. $\$9.80 \times 1.07 = \$10.486 \doteq \$10.49$
15. 48"
16. 96 square inches
17. $66\frac{2}{3}\%$ or 66.66...%
18. 0.6 to 0.3, or 0.6:0.3, or 2 to 1, or 2:1 or $\dfrac{0.6}{0.3}$ or $\dfrac{2}{1}$
19. 15 hours and 55 minutes
20. 2.03 pounds

from p. 169 — *fifth try*

1. 2.043
2. 21.98 inches
3. $0.08 \times 60 = 4.8$ inches
4. 4 yards – 3" = 144" – 3" = 141"
5. No. The sum of the digits equals 46, which is not evenly divisible by 3.
6. 0.9 ounces per minute
7. 61, 67.
8. 13 + 19 or 3 + 29. (1 + 31 will not work, because 1 is not a prime number.)
9. 439.23 square inches
10. (See diagram →)

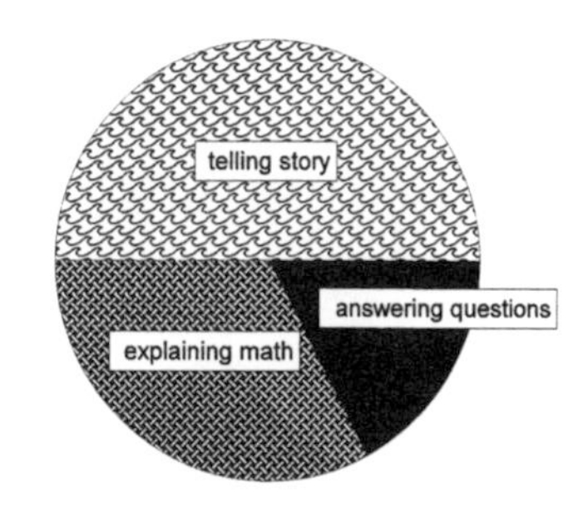

11. $\dfrac{1}{7} \times 161 = 23$ teachers in his class
12. $0.60 \times 97.3 = 58.38$ points
13. $\dfrac{0.81 \text{ pages}}{1 \text{ hour}} \times \dfrac{4.6 \text{ hours}}{1} = 3.726$ pages
14. $40 \times 1.10 = 44$ pages
15. 93.5 square inches
16. 43
17. 25% of the time had elapsed.
18. 7.2%
19. The missing digit must be 5 (so that the sum of the digits will be divisible by 9).
20. 7,000,000

from page 16

1. 87 = 4 score + 7 = **47**.

2. Look at a clock (12 hours). Look at a calendar (12 months). Look at a ruler (12 inches). Look at a jury (12 people). Look at eggs (dozen). Look at gold (12 troy ounces = 1 pound).

3. Did you ever wonder why there are sixty seconds in a minute, and sixty minutes in an hour? Now you know. In geometry we will study angles. A one-degree angle (written 1°) is very small. It takes 90 of them to make a right angle.

a right angle = 90°

1° angle

If you get out your microscope, and picture splitting a one-degree angle into 60 angles, each of those is called a minute. Sixty minutes of an angle equals one degree. If you take an angle that measures one minute and split it into 60 little angles, each of those would measure one second. These are super tiny angles.

4. $4\frac{2}{3} - 2\frac{3}{4} = 4\frac{8}{12} - 2\frac{9}{12} = 3\frac{12}{12} + \frac{8}{12} - 2\frac{9}{12} = 1\frac{11}{12}$

from page 18

1.
$$\begin{array}{r} 5.3 \\ 6.97 \\ \underline{8.888} \\ 21.158 \end{array}$$

2.
$$\begin{array}{r} 4.009 \\ 27.008 \\ \underline{9.} \\ 40.017 \end{array}$$

3.
$$\begin{array}{r} 270. \\ 2.99 \\ \underline{3.88} \\ 276.87 \end{array}$$

270 is the same as 270.

4. $\frac{14}{3} \times \frac{11}{4} = \frac{\overset{7}{\cancel{14}}}{3} \times \frac{11}{\underset{2}{\cancel{4}}} = \frac{77}{6} = 12\frac{5}{6}$

from page 21

1. 79.05000 = 79.05

(Please don't go crazy and turn 79.05 into 79.5. Those numbers aren't equal. 79.05 means 7 tens + 9 ones + 0 tenths + 5 hundredths, but 79.5 means 7 tens + 9 ones + 5 tenths.)

2.

$$\begin{array}{r} 3.070 \\ -\underline{1.008} \\ 2.062 \end{array}$$

3. ☒ yes (You *do* know that not everything you read in a book is true, don't you? But I still can't figure out why rong is spelled w-r-o-n-g.)

4. $4\frac{2}{3} \div 2\frac{3}{4} = \frac{14}{3} \div \frac{11}{4} = \frac{14}{3} \times \frac{4}{11} = \frac{56}{33} = 1\frac{23}{33}$

from page 23

1. 23.5 8880. 1. 7100.

2. 5000. 330. A million is 1,000,000. If you multiply by a thousand, you get 1,000,000,000, which is a billion.

3. The inverse function of *multiply by a thousand* is *divide by a thousand.* That undoes the original function.

For example, suppose I start with 32.91 and I multiply by a thousand. I'll get 32910. Then if I divide by a thousand (by moving the decimal to the *left* three places), I'll get back to 32.91. Magic! Isn't it?

4. The inverse function of *divide by ten* is *multiply by ten.*

5. In order to move the decimal of 0.6 three places to the left, I'll need to change 0.6 into 0000.6 first. Those numbers are equal.
0.6 means 0 ones + 6 tenths
0000.6 means 0 ten thousands + 0 thousands + 0 hundreds + 0 tens + 0 ones + 6 tenths.

The Double-Left Rule is *You can add zeros on the* left *of the last digit that is on the* left *of the decimal.*

6. 7.777 0.00002 You could write the answer to the third part in many different ways. You could write 0.0000 or 0 or 0. or 0.00000000.

7. 0.4 − 0.004 = 0.396

8. $6\frac{2}{5} + 2\frac{1}{2} = 6\frac{4}{10} + 2\frac{5}{10} = 8\frac{9}{10}$

9. A million is 1,000,000. For 10, you move the decimal one place to the right. For 100, you move it two. For 1000, you move it three.
For 1,000,000 you move it six places to the right.

10. A billion is 1,000,000,000. You would move the decimal nine places to the left.

11. One degree is equal to 60 minutes. So 7° would equal $7 \times 60 = 420$ minutes.

from page 27

1. As you recall, $\pi \approx 3.14159265358979$. The thousandths place is three decimal digits. So $3.141\not{5}9265358979 \doteq 3.142$
$\approx$ means *approximately equal to*

2. $0.8888 \doteq 0.9$
$3.90 \doteq 3.9$
$0.02 \doteq 0.0$
$\pi \doteq 3.1$
776. which equals 776.00000 rounds to 776.0 which could be written as 776. In other words, there was nothing to round, so it stayed the same.

3. $6\frac{2}{5} - 2\frac{1}{2} = 6\frac{4}{10} - 2\frac{5}{10} = 5\frac{10}{10} + \frac{4}{10} - 2\frac{5}{10} =$
$5\frac{14}{10} - 2\frac{5}{10} = 3\frac{9}{10}$

from page 36

1. 0.0546

2. 46.789

3. 0.00000009

4. 0.0006

5. 0. (or 0 or 0.0)

6. $6\frac{2}{5} \times 2\frac{1}{2} = \frac{32}{5} \times \frac{5}{2} = \frac{\cancel{32}^{16}}{\cancel{5}_{1}} \times \frac{\cancel{5}^{1}}{\cancel{2}_{1}} = 16$

7. 75 (or 75.)

from page 38

1. The inverse function of `start walking` is `stop walking`.

2. Will you always get the same answer? No. One time you might get Jeanette MacDonald, and another time you might get Nelson Eddy.

3. When I square 9 on Thursday, I get 81. When I square 9 on Saturday, I get 81. I always get 81 when I square 9. *Square the number* is a function.

4. Yes. If I start with any particular number, say, 34.657, and I round it to the nearest hundredths place, I will always get 34.66. I don't get 34.65 sometimes and 34.66 other times.

5. Suppose I tell you that I rounded a number to the nearest hundredth and got 34.66 as an answer. Can you tell me what the original number was? Can you undo this rounding? Can you say that 34.657 was the original number? No. It might have been 34.658, or it might have been 34.662, or 34.6600007. They all round to 34.66. *Rounding to the nearest hundredth* is a function that does not have an inverse.

Some functions have inverses and others don't. Getting a duck tattooed on your forehead is a function. But it doesn't have an inverse. There is no way to undo that and come back to your original, unscarred forehead.

One of the most famous is-there-an-inverse-function questions in all of literature came when the function *being born* was being discussed. A Jewish ruler named Nicodemus asked if that had an inverse function. "How can a man enter a second time into his mother's womb?"

6. $6\frac{2}{5} \div 2\frac{1}{2} = \frac{32}{5} \div \frac{5}{2} = \frac{32}{5} \times \frac{2}{5} = \frac{64}{25} = 2\frac{14}{25}$

7. I think multiplying will be easier. Multiplying the fractions would only involve multiplying top times top and bottom times bottom. Adding the fractions would involve about twice as many steps. Many people would have (mistakenly) guessed that addition was easier.

8. $C = \pi d \approx 3.14 \times 8 = 25.12$ feet

9. $8 \div 2 = 4$ feet

10. $7.7 - 6.09 = 1.61$

Your Turn to Play
COMPLETE SOLUTIONS

from page 41

1.

$$\begin{array}{r} \text{3 feet} \\ -\quad \text{1 inch} \\ \hline \end{array} \quad \textit{becomes} \quad \begin{array}{r} \text{2 feet 12 inches} \\ -\quad \text{1 inch} \\ \hline \text{2 feet 11 inches} \end{array}$$

2.

$$\begin{array}{r} \text{6 feet 8 inches} \\ -\ \text{5 feet 9 inches} \\ \hline \end{array} \quad \begin{array}{r} \text{5 feet 12 inches + 8 inches} \\ -\ \text{5 feet 9 inches} \\ \hline \end{array} \quad \begin{array}{r} \text{5 feet 20 inches} \\ -\ \text{5 feet 9 inches} \\ \hline \text{11 inches} \end{array}$$

Betty could wear an 11" hat.

3.

$$\begin{array}{r} \text{6 yards} \\ -\quad \text{1 inch} \\ \hline \end{array} \quad \textit{becomes} \quad \begin{array}{r} \text{5 yards 36 inches} \\ -\quad \text{1 inch} \\ \hline \text{5 yards 35 inches} \end{array}$$

The subtraction could also have been done by borrowing *twice*:

$$\begin{array}{r} \text{6 yards} \\ -\quad \text{1 inch} \\ \hline \end{array} \quad \begin{array}{r} \text{5 yards 3 feet} \\ -\quad \text{1 inch} \\ \hline \end{array} \quad \begin{array}{r} \text{5 yards 2 feet 12 inches} \\ -\quad \text{1 inch} \\ \hline \text{5 yards 2 feet 11 inches} \end{array}$$

4.

$$\begin{array}{r} \text{1 century} \\ -\quad \text{1 sec} \\ \hline \end{array} \quad \begin{array}{r} \text{99 years 365 days} \\ -\quad \text{1 sec} \\ \hline \end{array} \quad \begin{array}{r} \text{99 years 364 days 24 hours} \\ -\quad \text{1 sec} \\ \hline \end{array}$$

$$\begin{array}{r} \text{99 years 364 days 23 hours 60 minutes} \\ -\quad \text{1 sec} \\ \hline \end{array}$$

$$\begin{array}{r} \text{99 years 364 days 23 hours 59 minutes 60 seconds} \\ -\quad \text{1 second} \\ \hline \text{99 years 364 days 23 hours 59 minutes 59 seconds} \end{array}$$

from page 49

1. Because W = {0, 1, 2, 3, . . .}, you are correct if you wrote either 0 or 1 or 2 or 145982.

2. $A \cup B = \{\text{Sam, Pat, Chris, Joe}\}$, $A \cap B = \{\text{Pat}\}$, $A - B = \{\text{Sam, Chris}\}$.

3.

$$\begin{array}{r} \text{6 lbs. 1 oz.} \\ -\quad \text{2 oz.} \\ \hline \end{array} \quad \begin{array}{r} \text{5 lbs. 16 oz. + 1 oz.} \\ -\quad \text{2 oz.} \\ \hline \end{array} \quad \begin{array}{r} \text{5 lbs. 17 oz.} \\ -\quad \text{2 oz.} \\ \hline \text{5 lbs. 15 oz.} \end{array}$$

4. $9\frac{1}{2} + 5\frac{4}{5} = 9\frac{5}{10} + 5\frac{8}{10} = 14\frac{13}{10} = 15\frac{3}{10}$

from page 51

1. The digits add to 12, which is divisible by 3, so the number is also.

2. Multiplying by a billion will add nine zeros to the number, but it won't change the sum of the digits. So it will still be divisible by 3.

3. What does it mean to be evenly divisible by a number? It means that when you divide, you get a remainder of zero. If I want to know if 56 is evenly divisible by 8, I can divide and look at the remainder.

```
    7 R 0
8) 56
 – 56
    0
```

So what's the smallest whole number that is evenly divisible by 2, 3, 5, 17, and 39763? It is . . . zero. All these numbers divide evenly into zero with no remainder.

```
     0 R 0
17 ) 0
   – 0
     0
```

4. $9\frac{1}{2} - 5\frac{4}{5} = 9\frac{5}{10} - 5\frac{8}{10} = 8\frac{15}{10} - 5\frac{8}{10} = 3\frac{7}{10}$

from page 60

1.
```
      11.6
32) 371.2
  – 32
     51
  – 32
    192
  – 192
      0
```

Each family member would get 11.6 lbs. of meat.

2.
```
    156.18
7 ) 1093.26
  – 7
    39
  – 35
     43
  – 42
      12
    – 7
       56
     – 56
        0
```

$156.18 per person

3. $9\frac{1}{2} \times 5\frac{4}{5} = \frac{19}{2} \times \frac{29}{5} = \frac{551}{10} = 55\frac{1}{10}$ (or 55.1)

4. Because every natural number is a whole number, the natural numbers are a subset of the whole numbers.

5. $0.092 \times 0.007 = 0.000644$

6. $777.077 \doteq 777.$ ($\doteq$ means *rounded off to*)

7. $6^\circ \times \frac{60 \text{ minutes}}{1^\circ} = 360$ minutes.

The $\frac{60 \text{ minutes}}{1^\circ}$ is called a **conversion factor**. The numerator (the top) and the denominator (the bottom) are equal to each other. So the fraction is equal to one. And multiplying by one is always okay.

from page 63

1. $\begin{array}{r} 0.464 \\ 800\overline{)371.200} \end{array}$

2. 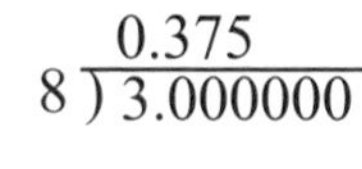

$\begin{array}{r} 0.375 \\ 8\overline{)3.000000} \end{array}$

3. $\begin{array}{r} 0.25 \\ 4\overline{)1.000000} \end{array}$

Now you know why \$0.25 is called a quarter ($\frac{1}{4}$).

> The word *liberty* is on every one of our coins. It doesn't just mean freedom.
> Liberty means freedom from being overgoverned.

4. $0.15 = \frac{15}{100}$ Because top and bottom are both evenly divisible by 5, we can reduce this fraction. $\frac{15}{100} = \frac{3}{20}$

5. $\frac{875}{1000}$ which, after a lot of reducing, equals $\frac{7}{8}$

6. $9\frac{1}{2} \div 5\frac{4}{5} = \frac{19}{2} \div \frac{29}{5} = \frac{19}{2} \times \frac{5}{29} = \frac{95}{58} = 1\frac{37}{58}$

from page 68

1. $\dfrac{0.07 \text{ lbs.}}{1} \times \dfrac{16 \text{ oz.}}{1 \text{ lb.}} = 1.12 \text{ oz.}$

2. $15 \overline{)\,0.4000}$ with quotient $0.026\overline{6}$ which rounds to 0.027 pounds.

3. $\dfrac{1}{7} = 0.142857142857142857142857142857142857\ldots = 0.\overline{142857}$

4. The logical consequence of those two given pieces of information is that π can never be expressed as a fraction.

5. $3\dfrac{1}{7} = 3.\overline{142857}$ (see problem 3 above). $3\dfrac{1}{7} > \pi$

> means *greater than.*

6. $2\dfrac{1}{3} + 4\dfrac{1}{7} = 2\dfrac{7}{21} + 4\dfrac{3}{21} = 6\dfrac{10}{21}$

7. The hard way to answer that question is to divide

$3 \overline{)\,20000000100060000}$ and see if it repeats.

The easy way is to remember that a number is evenly divisible by 3 if the sum of the digits is divisible by 3. The sum of the digits of 20000000100060000 is 9 and so 20000000100060000 is evenly divisible by 3. It won't be a repeating decimal.

from page 73

1. $0.25 \overline{)\,80.} \rightarrow 25 \overline{)\,8000.} \rightarrow 25 \overline{)\,8000.}$ with quotient 320

2. $80 \div \dfrac{1}{4} = \dfrac{80}{1} \div \dfrac{1}{4} = \dfrac{80}{1} \times \dfrac{4}{1} = \dfrac{320}{1} = 320$

3. $0.000001 \overline{)\,1000000} \rightarrow 1 \overline{)\,1000000000000.} \rightarrow 1 \overline{)\,1000000000000.}$ with quotient 1000000000000.

That's one, followed by 12 zeros. That's one trillion.

4. $4\dfrac{1}{7} - 2\dfrac{1}{3} = 4\dfrac{3}{21} - 2\dfrac{7}{21}$

$= 3\dfrac{21}{21} + \dfrac{3}{21} - 2\dfrac{7}{21}$

$= 3\dfrac{24}{21} - 2\dfrac{7}{21}$

$= 1\dfrac{17}{21}$

Some Big Numbers

one thousand	1,000
one million	1,000,000
one billion	1,000,000,000 = 10^9
one trillion	10^{12}
one quadrillion	10^{15}
one quintillion	10^{18}

10^9 means nine 10s multiplied together.

We'll learn about these little numerals ("exponents") in algebra.

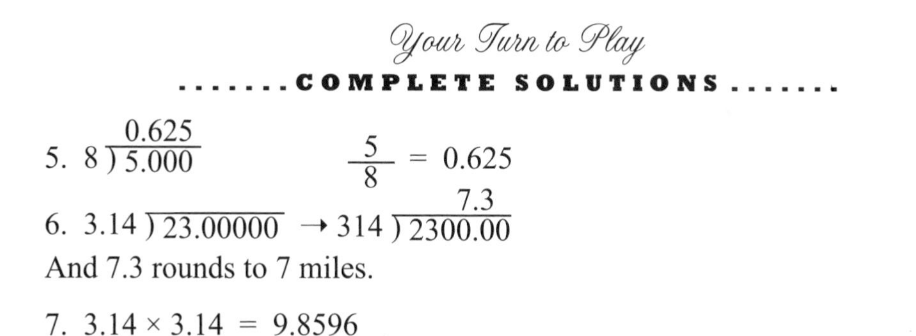

5. $8\overline{)5.000}$ with quotient 0.625 $\qquad \dfrac{5}{8} = 0.625$

6. $3.14\overline{)23.00000} \rightarrow 314\overline{)2300.00}$ with quotient 7.3

And 7.3 rounds to 7 miles.

7. $3.14 \times 3.14 = 9.8596$

from page 78

1. The numerical axis would go from 0 to either 2400 or, maybe, 2500.

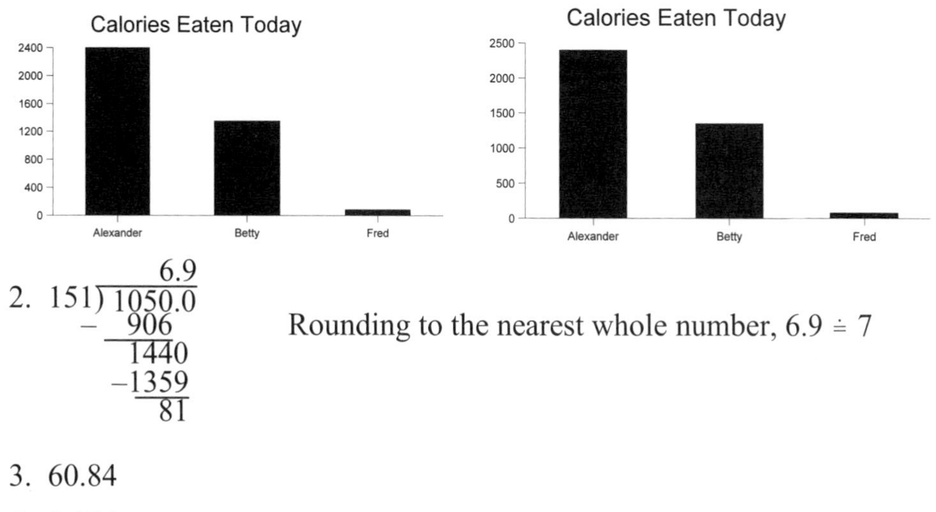

2. $151\overline{)1050.0}$ with quotient 6.9
$-\ 906$
1440
-1359
81

Rounding to the nearest whole number, $6.9 \doteq 7$

3. 60.84

4. 0.125

5. W – N = {0}. N – W = { }. { } is called the empty set.

from page 87

1. Three groups of 6. Six groups of 3. Nine groups of 2.

2. There is one natural number that has only one divisor. That is the number 1.

3. The even natural numbers are 2, 4, 6, 8, 10, 12. . . . The only one of them that is prime is 2. All the rest have more than two divisors.

4. 2, 3, 5, 7, 11, 13, 17, 19, 23, 29, 31, 37, 41, 43, 47, 53, 59.

5. The sum of the digits of 51 equals 6. Because 6 is divisible by 3, so is 51.

6. $4\frac{1}{7} \times 2\frac{1}{3} = \frac{29}{7} \times \frac{7}{3} = \frac{29}{\not{7}} \times \frac{\not{7}}{3} = \frac{29}{3} = 9\frac{2}{3}$

7. $1090 - 1.09 = 1088.91$

8. $25 + 26 + 27 + 28 + 29 = 135$.

9. For any two consecutive numbers, one of them has to be even. The even number will be composite.

from page 90 (INCOMPLETE SOLUTIONS)

1. $12 = 7 + 5 \quad 14 = 7 + 7 \quad 16 = 11 + 5$ etc.

2. $20 = 13 + 5 + 2$ etc. And we also don't know whether this second conjecture of Goldbach is true. Another open question. And another possibility for great fame.

from page 93

1. $A = \pi r^2 = 3.14 \times 45 \times 45 = 6358.5$ square feet.

2. $6358.5 \doteq 6359$ square feet. (Recall that $\doteq$ means "rounded off to.")

3. $A = s^2 = 10^2 = 10 \times 10 = 100$ square feet

4. The diameter of that circle would be 10 feet. Its radius would be 5 feet. The area of the circle would be $A = \pi r^2 = 25\pi$ square feet. If we approximate π by 3.14, then $25\pi \doteq 78.5$ square feet.

5. $A = \ell w = 78 \times 37 = 2886$ square miles

6. The largest circle that could be drawn inside that rectangle would have a diameter of 37 miles. Its radius would be 18.5 miles. The area of that circle would be $A = \pi r^2 = 3.14 \times 18.5 \times 18.5 = 1074.665$ square miles.

7. $1074.665 \doteq 1074.7$ square miles

8. $4\frac{1}{7} \div 2\frac{1}{3} = \frac{29}{7} \div \frac{7}{3} = \frac{29}{7} \times \frac{3}{7} = \frac{87}{49} = 1\frac{38}{49}$

9. If we wrote A = lw, it might look like we were writing 1w ("one w") since 1 ("one") and l ("L") look a lot alike in this font.

from page 96

1. She had confused dollars with cents. If two helicopters would cost a dollar, then each one would cost \$.50 or 50¢.

2. $.50¢ = \frac{50}{100}¢ = \frac{1}{2}¢$

3. $.50\overline{)100.} \rightarrow 50.\overset{200.}{\overline{)10000.}}$ You could buy 200 helicopters.

4. 42¢ 704¢ 80000¢ (or 80,000¢)

5. \$0.33 \$5.98 \$0.06

6. $A = \pi r^2 = 3.14 \times 8 \times 8 = 200.96$ square inches

7. $\$10.00 \div 200.96 \doteq \$0.049 \doteq \$0.05$ (or 5¢). Each square inch costs a nickel.

from page 98

1. The whole has to add up to 100%. Because Sunday school, taxes, and books add up to 90%, that leaves 10% for other things.

2. 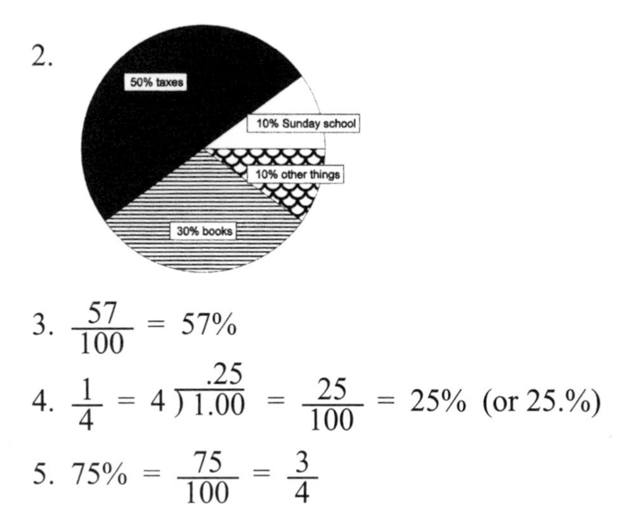

3. $\frac{57}{100} = 57\%$

4. $\frac{1}{4} = 4\overset{.25}{\overline{)1.00}} = \frac{25}{100} = 25\%$ (or 25.%)

5. $75\% = \frac{75}{100} = \frac{3}{4}$

6. $0.31 = \frac{31}{100} = 31\%$ (or 31.%)

You may have noticed that you really didn't have to first convert the decimal into a fraction. You can go directly from decimals to percents. Look how easy it is:

0.85 = 85.%
0.49 = 49.%
0.03 = 3.%

It is very similar to changing dollars into cents, as we did in the previous chapter. $0.85 = 85¢

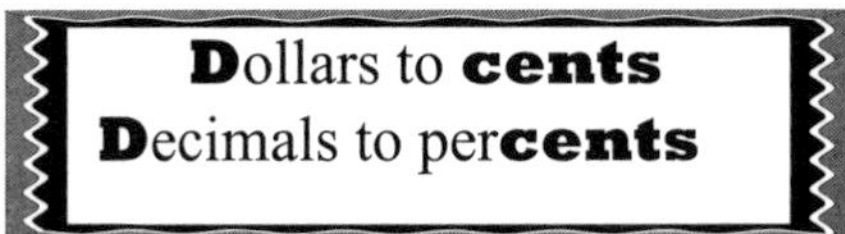

7. $3\frac{1}{12} + 7\frac{1}{4} = 3\frac{1}{12} + 7\frac{3}{12} = 10\frac{4}{12} = 10\frac{1}{3}$

8. $106.13\% = 1.0613$ (Move the decimal two places to the left.)

from page 105

1. 40% of 15 = $0.4 \times 15 = 6$ of them went out of business

2. 15% of 62.4 = $0.15 \times 62.4 = 9.36$ ounces of coconut

3. $7\frac{1}{12} - 3\frac{1}{4} = 7\frac{1}{12} - 3\frac{3}{12} = 6\frac{12}{12} + \frac{1}{12} - 3\frac{3}{12}$

$= 3\frac{10}{12} = 3\frac{5}{6}$

4. *Of* doesn't always mean multiply. When we sing, "My Country 'Tis of Thee," we don't multiply Country by God.

But when there are numbers on both sides of the *of*, then multiplication is indicated.

One-third of 414 = $\frac{1}{3}$ of 414 = $\frac{1}{3} \times 414 = \frac{1}{3} \times \frac{414}{1}$

$= \frac{414}{3} = 138$ verses that Joe made up.

5. $100\% - 9\% = 91\%$

from page 108

1. $\$71 \times 0.7 = \$49.70$
2. $\$41.31 \times \frac{2}{3} = \$27.54$
3. $\$20 \times \frac{2}{3} = \frac{40}{3}$ $\quad \frac{40}{3} \times \frac{3}{4} = 10$ $\quad 10 \times \frac{9}{10} = \$9$
4. $115 \times 0.9 = 103.5$ lbs.
5. $\$637.49 \times 0\% = \$0$
6. $84 \times 100\% = 84 \times 1.00 = 84$ students left

from page 110

1. $\$1.47$ per week $\times$ 52 weeks $= \$76.44$
2. 1 furlong per minute $\times$ 120 minutes $=$ 120 furlongs
3. $37.5 \times 96\% = 37.5 \times 0.96 = 36$ lbs.
4. $\frac{1}{8} = 8\overline{)1.000}$ (quotient $.125$) $= 12.5\%$
5. $7\frac{1}{12} \times 3\frac{1}{4} = \frac{85}{12} \times \frac{13}{4} = \frac{1105}{48} = 23\frac{1}{48}$

from page 113

1. $7 \times 132\% = 7 \times 1.32 = 9.24$ lbs.
2. $39 \times 108\% = 39 \times 1.08 = \$42.12$
3. $\$87 \times 0.97 = \$84.39$
4. We have mixed units of minutes and hours. First, change everything to minutes. 5 pages per minute $\times$ 120 minutes $=$ 600 pages.
5. $7 \times 1.42 = 9.94$ lbs.
6. $7 \times 58\% = 7 \times 0.58 = 4.06$ lbs.
7. 9 lbs. – 3 oz. = 8 lbs. + 16 oz. – 3 oz. = 8 lbs. 13 oz.
8. The subsets of S are: { }, {Star}, {Bird}, {Drum}, {Star, Bird}, {Star, Drum}, {Bird, Drum}, {Star, Bird, Drum}.

9. $56 \times 125\% = 56 \times 1.25 = 70$ shirts

10. $7\frac{1}{12} \div 3\frac{1}{4} = \frac{85}{12} \div \frac{13}{4} = \frac{85}{12} \times \frac{4}{13}$

$= \frac{85}{\cancel{12}_3} \times \frac{\cancel{4}^1}{13} = \frac{85}{39} = 2\frac{7}{39}$

from page 119

1. 726 square feet

2. Your work might look like: *First I tried 30 × 30 = 900. Way too small. Then I tried 90 × 90 = 8100. Too big. Then 50 × 50 = 2500. Too small. 70 × 70 = 4900. Getting close. 72 × 72 = 5184. Yes.*

3. A = (½)bh = ½ × 12 × 14 = 84 square inches

4. (½)bh = 0.5 × 7.8 × 6.9 = 26.91 square feet

5. We know that A = (½)bh. Putting in the numbers we know, we get 27 = ½ × b × 9. What is b? If you try out different guesses, it won't take you long to figure out that b = 6.

In algebra—which is coming up next—we will teach you how to solve problems like 27 = ½ × b × 9 without any guessing.

from page 128

1. The area of a parallelogram is base times height. The base is equal to 6 and the height is 8. The area is 48.

The area of a rectangle is length times width. The length is 8 and the width is 6. The area is 48. (Were you confused by the diagonal and the number 10 in the problem? Not every piece of information that is given in a problem needs to be used. Suppose you were given this problem: *Betty is driving a 3400-lb. Ford at 50 miles per hour for 3 hours. Her watch weighs 6 ounces and the temperature is 72°. How far did she drive?* Hopefully, you would say that she drove 150 miles.)

The base of the triangle is 8 and the altitude is 6, so the area is 24. (You also could have said that the base was 6 and the altitude was 8. You would have gotten the same answer.)

2. 50 miles per hour

3. As you work these trial-and-error square root problems, I hope you are noticing some shortcuts. We're trying to find $\sqrt{2809}$. We are looking for some number that when multiplied by itself will equal 280$\underline{9}$.

The last digit in your answer has to be either a 3 or a 7. When you square a number ending in 3 or 7, the answer will end in 9. For example,

$17^2 = 28\underline{9}$

$223^2 = 4972\underline{9}$

$8927513^2 = 7970048836516\underline{9}$

You know the answer will be between 50 ($50^2 = 2500$) and 60 ($60^2 = 3600$). So all you need to check is 53 and 57. $53^2 = 2809$, and you are done.

4. $9\frac{2}{7} + 2\frac{2}{5} = 9\frac{10}{35} + 2\frac{14}{35} = 11\frac{24}{35}$

5. $\frac{1}{3}$ $\quad 3\overline{)1.00}$ with quotient $0.33\frac{1}{3}$: $1.00 - 9 \to 10$, $10 - 9 \to 1$ $\quad = 33\frac{1}{3}\%$

6. $75\% = \frac{75}{100} = \frac{3}{4}$

from page 132

1. 75% of the height of the fence = 75% of 52" = 0.75×52 = 39"

2. 75% of the height of the fence = $\frac{3}{4}$ of 52" = $\frac{3}{4} \times \frac{52}{1}$

$= \frac{3}{\cancel{4}_1} \times \frac{\cancel{52}^{13}}{1} = 39"$

3. 26 is what percent of 52 ➞➞➞ $52\overline{)26.0}$ with quotient $.5$ $= 50\%$

4. $52 \times 1.4 = 72.8"$

5. 72.8" = 6' 0.8"

6. 2 is what percent of 3 ➞➞➞ $3\overline{)2.00}$ with quotient $.66\frac{2}{3}$ $= 66\frac{2}{3}\%$

7. $9\frac{2}{7} - 2\frac{3}{5} = 9\frac{10}{35} - 2\frac{21}{35} = 8\frac{35}{35} + \frac{10}{35} - 2\frac{21}{35} = 6\frac{24}{35}$

from page 135

1. The conversion factor is $\frac{\$8.61}{7 \text{ bookmarks}}$

$$\frac{11 \text{ }\cancel{\text{bookmarks}}}{1} \times \frac{\$8.61}{7 \text{ }\cancel{\text{bookmarks}}} = \frac{\$94.71}{7} = \$13.53$$

2. $\$8.61 \times 95\% = \$8.61 \times 0.95 = \$8.1795 \doteq \$8.18$

3. $\$8.61 \times \frac{2}{3} = \frac{8.61}{1} \times \frac{2}{3} = \frac{17.22}{3} = \$5.74$

4. There are several ways you could give your answer:

   i) 8 women to 6 men
   ii) 8:6
   iii) $\frac{8 \text{ women}}{6 \text{ men}}$
   iv) $\frac{4 \text{ women}}{3 \text{ men}}$ (reducing the fraction)
   v) 4:3

5. $9\frac{2}{7} \times 2\frac{3}{5} = \frac{65}{7} \times \frac{13}{5} = \frac{\overset{13}{\cancel{65}}}{7} \times \frac{13}{\underset{1}{\cancel{5}}} = \frac{169}{7} = 24\frac{1}{7}$

from page 139

1. Stanford is mapped to no.

2. The rule could be *does Fred teach there?* It could be *is the school spelled with all capital letters?* It could be *is this school in Kansas?* There can be more than one rule that describes a function.

3. Sacramento → California, Boise → Idaho, Dover → Delaware

4. {(10, 3), (52, 45), (7,0)}

5. If it had an inverse, then *no* would be mapped to several different images. That's the one thing that can't happen with a function. Each thing must have exactly one image. You can't have no → Harvard and also have no → Stanford.

6. $9\frac{2}{7} \div 2\frac{3}{5} = \frac{65}{7} \div \frac{13}{5} = \frac{65}{7} \times \frac{5}{13} = \frac{\overset{5}{\cancel{65}}}{7} \times \frac{5}{\underset{1}{\cancel{13}}} = \frac{25}{7} = 3\frac{4}{7}$

7. The rule is *how many syllables does it have?*

Your Turn to Play

.......COMPLETE SOLUTIONS.......

from page 143

1. The third graph contained the points (2, 1) and (2, 3). The number 2 was mapped to 1 and also to 3. The definition of function is that each first coordinate can have only one image.

2. If you see two points—one directly over the other—then it is not a function.

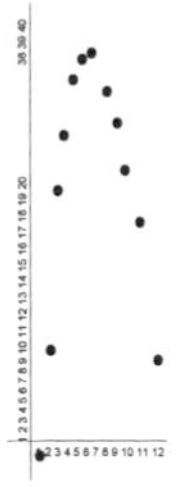

3. The graph would get pretty scrunched up if you put the same scale on both axes.

4. –8° is seven degrees colder than –1°.

If it were –8° in your bedroom and you turn on the heat, it would take an eight-degree rise in temperature to get you up to 0°. It would only take a one-degree rise in temperature to go from –1° up to 0°. In algebra, we write $-8 < -1$.

You know it's below zero degrees in your bedroom if you fall out of bed and break your pajamas.

from page 153

1. 33⅓%, 62½%, 25%, 50%, 12½%

2. 66⅔%, 87½%, 37½%, 75%

from page 156

1. 11:48 A.M. to noon is 12 minutes. Noon to 12:26 P.M. is 26 minutes.
   $12 + 26 = 38$ minutes

2. One-eighth of an ounce is 0.125 ounces. (You know that because $\frac{1}{8}$ = 12.5%.) $0.125 - 0.08 = 0.045$ ounces were used from the new tube.

3. $8\frac{1}{3} + 1\frac{4}{5} = 8\frac{5}{15} + 1\frac{12}{15} = 9\frac{17}{15} = 10\frac{2}{15}$

4. From 12:10 P.M. on one day to 12:10 P.M. two days later is 48 hours. Then from 12:10 P.M. to 1 P.M. is 50 minutes which is $\frac{5}{6}$ of an hour. Then from 1 P.M. to 5 P.M. is 4 hours. $48 + \frac{5}{6} + 4 = 52\frac{5}{6}$ hours.

5. {(Fred, Alexander), (Peggy, Alexander), (Betty, Alexander), (Alexander, Alexander)}

from page 159

1. Three out of 24 is $\frac{3}{24}$ or $\frac{1}{8}$ which equals 12½%.

2. There are 2 chances out of 3 that the student will be female. That's $\frac{2}{3}$ or 66⅔%. (These questions are a lot easier if you have memorized the Nine Conversions.)

3. The smallest probability is 0%. It occurs when something is certain not to happen. There are many examples of things with zero probability, and your example will differ from mine. My example of an event with zero probability is: selecting one of the 24 students in Fred's set theory class, who owns a tapir that likes to quote Cecil Rodd's famous advertising slogan for Wall's ice cream, which Rodd wrote in the spring of 1922.

I always say,
"Stop me and buy one."

There is virtually zero probability that a reader of *Life of Fred: Decimals and Percents* thought of the example I have just given.

I did! . . . Just kidding.

4. Nothing can be more than absolutely certain, and that is a probability of 100%.

5. $8\frac{1}{3} \times 1\frac{4}{5} = \frac{25}{3} \times \frac{9}{5} = \frac{\overset{5}{\cancel{25}}}{\underset{1}{\cancel{3}}} \times \frac{\overset{3}{\cancel{9}}}{\underset{1}{\cancel{5}}} = 15$

6. From 8:12 A.M to 12:12 P.M. is 4 hours. From 12:12 P.M. to 3:12 P.M. is 3 hours. From 3:12 P.M. to 3:41 P.M. is 29 minutes.

4 hrs. + 3 hrs. + 29 mins. = 7 hrs. 29 mins.

Index

What Comes after Decimals and Percents?

THREE MIDDLE SCHOOL BOOKS
BEFORE
HIGH SCHOOL MATH

① *Life of Fred: Pre-Algebra 0 with Physics*
② *Life of Fred: Pre-Algebra 1 with Biology*
③ *Life of Fred: Pre-Algebra 2 with Economics*

These three books are the three links to take you from decimals and percents up to beginning algebra. Each book builds on the things learned in the previous book. (Translation: Don't skip any of these.)

In addition to getting you ready for high school algebra . . .

In the *Physics* book you will also learn a lot of physics.
In the *Biology* book you will also learn a lot of biology.
In the *Economics* book you will also learn a lot of economics.

To see what other books
have been written
about Fred
please visit

FredGauss.com